DESIGN OF UPLIFT-RESISTING ANCHORS FOR SHIPS AND SUBMARINES

DEEP OCEAN TECHNOLOGY

by

R.J. Taylor
D. Jones
and
R.M. Beard

Wexford Press
2008

CONTENTS

	page
Chapter 1. INTRODUCTION	1
1.1. Purpose of Handbook	1
1.2. Background of Uplift-Resisting Anchors	1
Chapter 2. OPERATIVE TYPES	3
2.1. Propellant-Actuated Direct-Embedment Anchors	3
2.2. Vibrated Direct-Embedment Anchors	6
2.3. Screw-In Anchors	7
2.4. Driven Anchors	7
2.5. Drilled Anchors	9
2.6. Deadweight Anchors	9
Chapter 3. DATA SUMMARIES	11
3.1. Magnavox Embedment Anchor System, Model 1000 (Propellant-Actuated)	13
3.2. Magnavox Embedment Anchor System, Model 2000 (Propellant-Actuated)	19
3.3. VERTOHOLD Embedment Anchor, 10K (Propellant-Actuated)	25
3.4. SEASTAPLE Explosive Embedment Anchor, Mark 5 (Propellant-Actuated)	29
3.5. SEASTAPLE Explosive Embedment Anchor, Mark 50 (Propellant-Actuated)	33
3.6. CEL 20K Propellant Anchor (Propellant-Actuated)	37
3.7. CEL 100K Propellant Anchor (Propellant-Actuated)	43
3.8. Explosive Embedment Anchor, XM-50 (Propellant-Actuated)	45
3.9. Explosive Embedment Anchor, XM-200 (Propellant-Actuated)	47
3.10. PACAN 3DT (Propellant-Actuated)	51
3.11. PACAN 10DT (Propellant-Actuated)	55
3.12. Direct-Embedment Vibratory Anchor (Vibrated)	57
3.13. Vibratory Embedment Anchor, Model 2000 (Vibrated)	63
3.14. Chance Special Offshore Multi-Helix Screw Anchor (Screw-In)	67
3.15. Stake Pile (Driven)	71
3.16. Umbrella Pile-Anchor, Mark III (Driven)	75
3.17. Umbrella Pile-Anchor, Mark IV (Driven)	79
3.18. Rotating Plate Anchor (Driven)	83
3.19. Expanded Rock Anchor (Drilled)	85
3.20. Free-Fall Anchor System (Deadweight)	87
Chapter 4. OTHER PROSPECTIVE TYPES	93
4.1. Implosive Anchor	93
4.2. Free-Fall Anchor	93
4.3. Pulse-Jet Anchor	95
4.4. PADLOCK Anchor System	97
4.5. Jetted-In Anchor	101
4.6. Hydrostatic Anchor	102
4.7. Seafloor Rock Fasteners	102

	page
Chapter 5. APPLICABLE COMPUTATIONS	107
5.1. Penetration	107
5.2. Holding Capacity	111
5.3. Sample Problem	118
Chapter 6. REFERENCES, BIBLIOGRAPHY, AND PATENTS	121
6.1. References	121
6.2. Bibliography	122
6.3. Patents	128
APPENDIXES	
A – Supplementary Tabular Data on Specific Anchor Designs	129
B – Curves for Short-Term Static Holding Capacity Versus Depth	139
C – Nomographs for Calculating Holding Capacity	147

LIST OF ILLUSTRATIONS

Figure 2.1-1.	General configuration of a propellant-actuated anchor.	4
Figure 2.1-2.	Embedment and keying of a propellant-actuated anchor.	4
Figure 2.1-3.	Anchor-projectile with hinged flukes extended.	4
Figure 2.1-4.	Three-finned anchor-projectile for coral seafloor.	4
Figure 2.2-1.	Deep-water vibrated anchor.	6
Figure 2.4-1.	Driven anchors.	8
Figure 2.6-1.	Primitive deadweight anchor.	9
Figure 3.1-1.	Magnavox embedment anchor system, Model 1000; cutaway view of anchor-projectile and gun assembly mounted in expendable gun assembly.	15
Figure 3.1-2.	Magnavox embedment anchor system, Model 1000; without deployment canister.	16
Figure 3.1-3.	Magnavox embedment anchor system; anchor-projectiles for Model 1000 (right) and Model 2000 (left).	17
Figure 3.2-1.	Magnavox embedment anchor system, Model 2000; reusable gun assembly and accessories	21
Figure 3.2-2.	Magnavox embedment anchor system, Model 2000; without reaction cone and gun stand assembly.	22
Figure 3.2-3.	Magnavox embedment anchor system, Model 2000; anchor with flukes fully opened	23
Figure 3.3-1.	VERTOHOLD embedment anchor; flukes fully opened.	27
Figure 3.3-2.	VERTOHOLD embedment anchor; flukes in penetrating position.	27
Figure 3.3-3.	VERTOHOLD anchor assembly rigged for command-firing and recovery of gun assembly	28
Figure 3.3-4.	VERTOHOLD anchor assembly rigged for automatic-firing and nonrecovery of gun assembly.	28
Figure 3.4-1.	SEASTAPLE embedment anchor, Mark 5.	30
Figure 3.4-2.	SEASTAPLE embedment anchor, Mark 5; rigged for recovery of gun assembly	31
Figure 3.5-1.	SEASTAPLE embedment anchor, Mark 50.	35
Figure 3.6-1.	CEL 20K Propellant Anchor.	39
Figure 3.6-2.	Schematic of CEL 20K Propellant Anchor.	40
Figure 3.6-3.	CEL 20K Propellant Anchor; rock fluke and piston.	41
Figure 3.6-4.	CEL 20K Propellant Anchor; sand fluke and piston in penetrating position.	42
Figure 3.6-5.	CEL 20K Propellant Anchor; sand fluke and piston in keyed position.	42

page

Figure 3.7-1. CEL 100K Propellant Anchor; launch vehicle with dummy anchor (used to evaluate gun performance). 44
Figure 3.7-2. CEL 100K Propellant Anchor; flukes. 44
Figure 3.8-1. Army explosive embedment anchor, XM-50; front quarter view 46
Figure 3.8-2. Army explosive embedment anchor, XM-50; rear quarter view 46
Figure 3.9-1. Army explosive embedment anchor, XM-200; cutaway model 49
Figure 3.9-2. Army explosive embedment anchor, XM-200; front quarter view 50
Figure 3.9-3. Army explosive embedment anchor, XM-200; rear quarter view 50
Figure 3.10-1. PACAN 3DT; equipped with plate-type fluke mounted in cradle aboard ship 52
Figure 3.10-2. PACAN 3DT; anchor-projectiles for sediments 53
Figure 3.10-3. PACAN 3DT; anchor-projectiles 53
Figure 3.11-1. PACAN 10DT; without anchor projectile 56
Figure 3.12-1. Navy vibratory anchor 59
Figure 3.12-2. Navy vibratory anchor; quick-keying fluke shown in position assumed after keying . . 60
Figure 3.12-3. Navy vibratory anchor; fluke locking mechanism 61
Figure 3.12-4. Navy vibratory anchor; embedded in sand on beach to demonstrate collapsible support-guidance frame 62
Figure 3.13-1. Ocean Science and Engineering vibratory embedment anchor, Model 2000 65
Figure 3.14-1. Chance Special Offshore Multi-Helix system for pipeline anchoring; pipeline bracket visible 68
Figure 3.14-2. Chance Multi-Helix screw anchor 69
Figure 3.14-3. Chance Special Offshore Multi-Helix system for pipeline anchoring 70
Figure 3.15-1. Navy stake pile; 8-inch. 72
Figure 3.15-2. Design specifications for Navy stake pile 73
Figure 3.16-1. Navy umbrella pile-anchor, Mark III; after test in sand 76
Figure 3.16-2. Navy umbrella pile-anchor, Mark III. 77
Figure 3.17-1. Navy umbrella pile-anchor, Mark IV. 80
Figure 3.17-2. Navy umbrella pile-anchor, Mark IV; after test in sand 81
Figure 3.18-1. Ménard rotating plate anchor. 84
Figure 3.19-1. Ménard expanded rock anchor; placement of chain into drilled hole 86
Figure 3.20-1. Delco free-fall anchor (12,000-pound anchor) 89
Figure 3.20-2. Delco free-fall anchor (typical) 89
Figure 3.20-3. Delco free-fall anchor (24,000-pound); mounted on launching platform on USCGS Rockaway 90
Figure 3.20-4. Delco free-fall anchor in typical deep-water mooring system 91
Figure 4.1-1. Propelled-shaft embedment of implosive anchor (Rossfelder and Cheung, 1973) . . . 94
Figure 4.1-2. Propelled-casing embedment of implosive anchor (Rossfelder and Cheung, 1973) . . 96
Figure 4.2-1. Free-fall embedment anchor 97
Figure 4.3-1. Mass drag reactor of the Pulse-Jet Anchor System (Lair, 1967) 98
Figure 4.3-2. Ballistic embedding anchor of the Pulse-Jet Anchor System (Lair, 1967) 98
Figure 4.4-1. Basic concept of PADLOCK Anchor System (Dantz, 1968). 99
Figure 4.4-2. PADLOCK Anchor System developed for test and evaluation (Dantz, 1968) . . . 100
Figure 4.5-1. Illustration of Jetted Anchor 101
Figure 4.6-1. Schematic of the Hydrostatic Anchor 102
Figure 4.7-1. Drive-set rock bolt; slot and wedge type (Brackett and Parisi, 1975) 103
Figure 4.7-2. Drive-set rock bolt; cone and stud anchor type (Brackett and Parisi, 1975) . . . 103
Figure 4.7-3. Torque-set rock bolt (typical) (Brackett and Parisi, 1975) 103
Figure 5.1-1. Incremental calculation flow for momentum penetration in clay and sand 109
Figure 5.1-2. Theoretical bearing capacity factor, N_q, versus angle of internal friction, ϕ, for a strip foundation 111

	page
Figure 5.2-1. Prediction procedure for direct-embedment anchor holding capacity	113
Figure 5.2-2. Reduction factor to be applied to field anchor tests in cohesive soils to account for suction effects	115
Figure 5.2-3. Plot for calculating D_c, the distance above the anchor, at which the characteristic strength, C_a, is to be taken	116
Figure 5.2-4. Design curves of holding capacity factor, $\bar{N}c$, versus relative embedment depth (D/B)	117
Figure 5.2-5. Recommended properties for a hypothetical cohesive soil when data on actual cohesive soil are not available	118
Figure 5.2-6. Holding capacity factor, $\bar{N}_q$, versus relative depth for cohesionless soil, c = 0	118
Figure 5.3-1. Vane shear strength profile for sample problem	119
Figure B-1. Short-term static holding capacity versus depth for small uplift-resisting anchors embedded in the cohesive soil described by Figure 5.2-5	140
Figure B-2. Short-term static holding capacity versus depth for intermediate uplift-resisting anchors embedded in the cohesive soil described by Figure 5.2-5	141
Figure B-3. Short-term static holding capacity versus depth for large uplift-resisting anchors embedded in the cohesive soil described by Figure 5.2-5	142
Figure B-4. Holding capacity versus depth for small uplift-resisting anchors embedded in the sand described by $\phi = 30°$ and $\gamma_b = 60$ pcf	143
Figure B-5. Holding capacity versus depth for intermediate uplift-resisting anchors embedded in the sand described by $\phi = 30°$ and $\gamma_b = 60$ pcf	144
Figure B-6. Holding capacity versus depth for large uplift-resisting anchors embedded in the sand described by $\phi = 30°$ and $\gamma_b = 60$ pcf	145
Figure C-1. Nomograph for calculating short-term holding capacity in cohesive soil in the 0-to-10-kip range	148
Figure C-2. Nomograph for calculating short-term holding capacity in cohesive soil in the 0-to-50-kip range	149
Figure C-3. Nomograph for calculating short-term holding capacity in cohesive soil in the 0-to-200-kip range	150
Figure C-4. Nomograph for calculating holding capacity in sand in the 0-to-10-kip range	151
Figure C-5. Nomograph for calculating holding capacity in sand in the 0-to-100-kip range	152
Figure C-6. Nomograph for calculating holding capacity in sand in the 0-to-300-kip range	153

LIST OF TABLES

Table 4.7-1. Parameters Affecting Holding Strength of Seafloor Fasteners (From Brackett and Parisi, 1975)	104
Table 5.1-1. Values of Side Adhesion Factor, δ, at High Velocity Derived From Field Test Data	108
Table 5.1-2. Ratio of Keying Distance to Fluke Length	111
Table A-1. Summary of Characteristics of Uplift-Resisting Anchors	131
Table A-2. Holding Capacity Data	135
Table A-3. Test Penetration Data	137

ACKNOWLEDGMENT

Grateful appreciation is extended to Mr. J. E. Smith for his detailed review and editing of the handbook and for his assistance in organizing the handbook into a presentable format.

Chapter 1. INTRODUCTION

1.1. PURPOSE OF HANDBOOK

The purpose of this handbook is to (1) identify and document the status of special types of anchors having the capability to resist uplift forces; (2) provide data on the properties and performance of these special anchors; (3) consolidate the data in order to facilitate anchor selection; and (4) establish a reference that can be readily updated to incorporate new data and new developments. Descriptions and data on anchors that are currently either shelf items or in an advanced stage of development are presented. Also, information on other less advanced designs and concepts is given. Sizes, weights, and operational characteristics of these special anchors, plus methods for estimating their penetration into seafloor sediments and their pull-out resistance, are provided.

This handbook includes material and information that was possible to obtain within a specified time frame. The development of embedment anchors continues, and additional information will be incorporated as it becomes available.

1.2. BACKGROUND OF UPLIFT-RESISTING ANCHORS

As ocean operations and construction have expanded and moved to deeper waters, the need for more sophisticated anchoring systems has emerged. A particular need is for anchors that can resist uplift and are highly efficient, reliable, and light weight where practicable. Other qualities desired are simplicity in handling and the facility for rapid installation.

Anchors that can resist uplift can significantly reduce the scopes of line associated with conventional drag anchors and also the quantity and sizes of accessories. They minimize the need for multileg arrangements to limit watch circle and lessen load-handling equipment requirements. They typically can be installed directly into the seafloor without the necessity of dragging, thus simplifying installation and improving positioning accuracy. They can sustain lateral as well as uplift loading. They broaden the range of feasible anchoring sites, such as on sloping and rocky seafloors, that are considered to be off limits with conventional anchors. They potentially can significantly reduce lowering and placement times, thus making ocean operations less vulnerable to adverse sea and weather conditions.

In deep water, cost efficiency can become the primary reason for utilizing anchors that resist uplift because installation time and line scope become increasingly significant factors as water depth increases. In shallow water, particularly in well-used harbors, uplift-resisting anchors have the advantage of eliminating considerable bottom gear that can be damaged by ship anchors.

Until recently, only a limited selection of anchor types, which were comprised mainly of conventional drag anchors, deadweight anchors, and piles, were available when designing an anchoring system to resist uplift loading. Conventional drag anchors are inefficient for this mode of loading, because they rely principally on their own weight plus that of the sinkers which ensures lateral loading on the drag anchor. Deadweight anchors are heavy to transport and handle for the effective holding to be gained. They are susceptible to drifting, and they are unreliable on sloping seafloors. Piles are limited presently to relatively shallow water.

Commencing in the 1960s numerous anchor concepts were proposed that could counter uplift loading. They included a variety of types, such as propellant-actuated, vibrated, screw-in, implosive, pulse-jet, jetted, and hydrostatic. Some advanced to the development stage, encountered problems, and were abandoned. Others have demonstrated potential workability, but require additional validation testing. A few have been developed to the point of being considered operational hardware.

Despite the progress that has been made with the new anchor concepts, some difficulties remain. Seafloors with anomalous conditions — such as shallow sediment over rock; weathered and fractured rocks; seafloors with gravel and boulders interspersed; and seafloors layered by turbidities — make penetration of the seafloor uncertain and the prediction of holding capacity unreliable. Where seafloor slopes are greater than 10%, the orientation of special anchors for proper penetration is difficult and uncertain.

Deep-water techniques for anchoring in rock are limited to drilled-in piles. Less expensive, more controllable and rapid procedures are needed. Still more expedient means to install all deep-ocean anchor systems are needed. Future installations will impose even more severe anchor requirements. Anchoring systems with 100-to-1,000-kips holding capacities are envisioned in deep water. Multiple or modular anchors and piles are a potential solution, but knowledge of their interaction and resulting performance must be gained for them to become practical. Anchoring technology is being advanced to meet these challenges.

Chapter 2. OPERATIVE TYPES

Anchors designed to resist uplift are separated into the following categories:

- Propellant-acutated direct-embedment anchors
- Vibrated direct-embedment anchors
- Screw-in anchors
- Driven anchors
- Drilled anchors
- Deadweight anchors
- Free-fall anchors

Each category of anchor is described, and distinguishing characteristics are identified. Modes of operating, handling, and placing the anchor are given. Advantages and disadvantages are listed. Also, a brief history and the current status of the anchor are summarized. Information on operative designs is given in Section 3.

2.1. PROPELLANT-ACTUATED DIRECT-EMBEDMENT ANCHORS

2.1.1. Description

A propellant-actuated anchor (often referred to as an explosive anchor) is one that is propelled directly into the seafloor at a high velocity by a gun. Basically, it consists of an anchor-projectile and a gun assembly comprised of a gun and a reaction vessel. Though a variety of forms has evolved, Figure 2.1-1 illustrates the general design of such anchors. The anchor-projectile includes a piston and fluke. The gun incorporates a safe-and-arm device that is actuated by hydrostatic pressure, which arms the gun only after a predetermined depth is attained. A propellant charge, contained in a cartridge, generates the gas pressures that accelerate the anchor-projectile into the seafloor. Whenever possible the gun assembly is recovered and used again. Recovery becomes increasingly difficult at depths greater than 1,000 feet.

There are presently two types of projectiles* for use in sediments. In the first type, the portion which engages the soil to resist pull-out (the fluke) is a rotating plate assembly. It can be either a single-plate construction or a trihedron construction of three flat plates (Y fluke). The plates enter the seabed edgewise. After emplacement, an upward pull on the anchor line, which is transmitted to the fluke at an eccentric connection, "keys" the fluke; that is, the fluke rotates to a position in which maximum bearing area is presented to the soil to resist pull-out. Figure 2.1-2 illustrates the plate-like fluke and the "keying" action.

In the second type of sediment projectile, two or more slender, movable flukes are hinged to the cylindrical body of the projectile. During penetration of the seafloor, they are clustered tightly about the body. Then, when a load is applied, they key by opening outward. Figure 2.1-3 shows this type of sediment projectile in the open position.

Existing projectiles for coral and rock do not have flukes. The projectile is shaped like a spear or arrowhead to achieve maximum penetration, and the lateral surfaces that engage the surrounding material can be serrated. Projectiles for use in coral and rock include a solid shaft with hardened point and serrated neck, a flat arrowhead shape, and a "three-dimensional" arrowhead (a pointed trihedron of flat plates with serrated or nonserrated edges). Figure 2.1-4 shows a coral rock type of anchor projectile.

---

* For convenience the term "projectile" is sometimes used for "anchor-projectile."

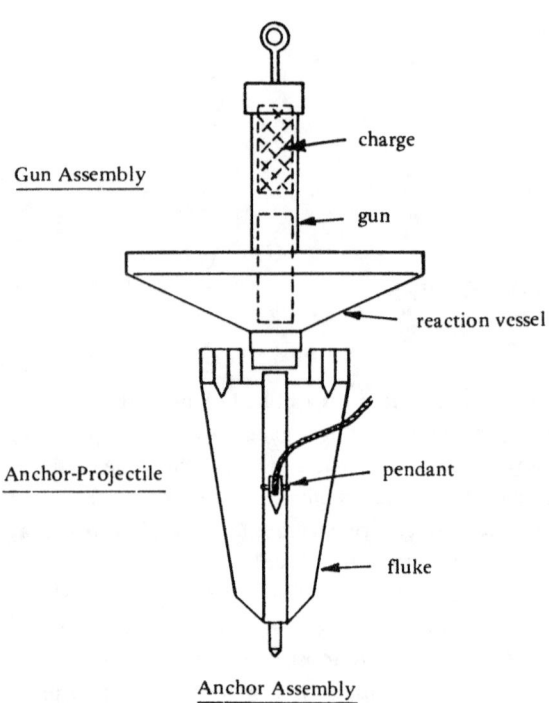

Figure 2.1-1. General configuration of a propellant-actuated anchor.

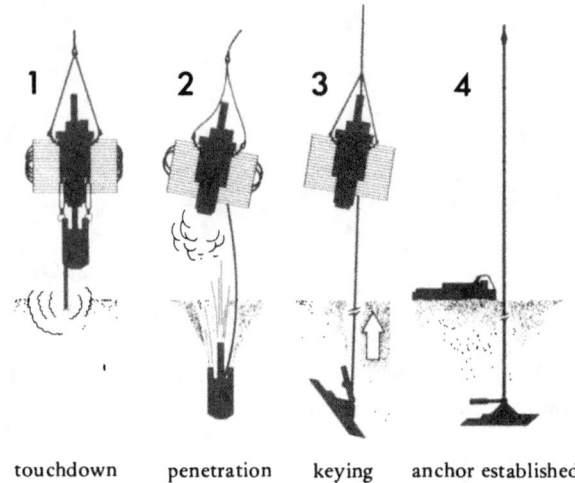

Figure 2.1-2. Embedment and keying of a propellant-actuated anchor.

Figure 2.1-3. Anchor-projectile with hinged flukes extended.

Figure 2.1-4. Three-finned anchor-projectile for coral seafloor.

The reaction vessel can be configured into practically any form that provides effective mass and high drag to minimize recoil and ensure optimum projectile velocity. It can be designed to entrap water to provide the mass (drag cones or plates are examples), or it can simply utilize the mass of the steel. The latter approach is less costly, but the resulting anchor system is heavier and recoil distances are greater. Reaction distances can vary from about 8 to 50 feet depending upon the reaction vessel configuration and the effective mass.

Several techniques are available for placing and firing propellant-actuated anchors. Such factors as the size and design of anchor, depth of water, handling equipment, and the overall operational requirements dictate the method to be used. The anchor can be fired by lowering it until a probe extending below it touches the bottom and triggers the firing mechanism. Or, the anchor can be held suspended above the bottom and fired by a signal from the surface through a firing line. The latter method is limited to depths less than about 200 feet and requires close control of movement of the surface work platform. A third method is to position the anchor on the seafloor by means of a support frame. In one design the reaction vessel also serves as a support frame. In this case, the anchor is properly oriented to fire the anchor perpendicular to the bottom without regard for bottom slope. Firing the anchor with a support frame is usually achieved by signal through a firing line from the surface. However, coded sound signals for firing the anchor are possible. Also, a touchdown sensor with a delay mechanism that permits the anchor to attain its proper orientation on the bottom before ignition has been used successfully.

Depending upon the mode of operation for a propellant-actuated anchor, up to three cables from the surface may be needed — the main anchor line, a line for retrieving the gun assembly, and an electrical cable for remote firing of the gun. In water less than 600 feet, two or more lines can be lowered without entanglement if proper precautions are taken. One line is attached to the gun assembly and the other to the anchor. The firing cable can be a separate line or be attached to or incorporated with the gun assembly line. After firing the gun assembly is retrieved. In deep water, only a single line can be lowered; as a result, the gun assembly is usually considered expendable. Another alternative in deep water is to free fall the anchor with the anchoring line stored in a bale on the anchor. A novel approach for retrieving the gun assembly has been developed by S. N. Marep*. A single line is attached to the gun assembly during lowering and firing. After firing, the gun assembly is retrieved, and a small diameter line, which is attached to the anchor downhaul cable (a short cable attached to the fluke) and gun assembly and located in the reaction case, is unreeled. The main anchoring line is placed over this guideline and lowered until it clamps to the downhaul cable. This technique is usable to depths as great as 3,000 feet.

2.1.2. Advantages and Disadvantages

The principal advantages of propellant-actuated anchors are: (1) the anchor assembly is a compact package and has a higher holding capacity/weight efficiency than other anchors of the same capacity. (2) The anchor can function in a broad range of sediments and in material at least as hard as coral and vesicular basalt. (3) The concept is very nearly perfected. (4) Because penetration is rapid, special efforts to keep the surface vessel on station during embedment of the anchor are not required. (5) The light weight simplifies operational and handling difficulties.

The principal disadvantages are: (1) This type of anchor is not suited for a seafloor where there is rubble, medium-to-large-size boulders, pillow basalt, or rock overlain by sediment. (2) Special shipment, storage, and handling is required for the ordnance features. (3) The gun assembly is not generally retrievable in deep water. (4) The downhaul cable that subsequently becomes part of the anchor line is susceptible to abrasion and deterioration.

2.1.3. History and Status

Propellant anchors were first developed in the late 1950s. Since then anchors ranging in nominal holding capacity from 1,000 to 220,000 pounds have been developed and tested. Most of the anchors were designed for shallow-water applications (less than 600 feet), but some can be used in depths of more than 10,000 feet and a few were designed for operation to 20,000 feet.

---

* The developer of the PAGAN anchors, Sections 3.10 and 3.11.

Propellant-actuated anchors are still basically in their infancy. Considerably more testing and actual field use are required to develop user confidence in their unique capabilities and to eliminate the onus of fear and uncertainty that surrounds them.

2.2. VIBRATED DIRECT-EMBEDMENT ANCHORS

2.2.1. Description

A vibrated anchor (also referred to as "vibratory" anchor) is one that is driven into the seafloor by vibration. It is a long, slender metal construction consisting of a fluke-shaft assembly and a vibrator; for deep-water use (greater than 600 feet), a support guidance frame and a storage battery power pack are required. The deep-water system is illustrated in Figure 2.2-1; the shallow-water system is shown in Section 3.13.

The vibrator that drives this type of anchor consists of counter-rotating eccentric masses* which can be either hydraulically driven from the surface or electrically powered at the seafloor.

The fluke used for both the Ocean Science and Engineering (OSE) anchor (shallow water) and the CEL anchor (deep water) is the special rotating Y-fluke developed under the CEL free-fall anchor program (Smith 1966). A variety of sizes is available for the anchors. It has been shown both analytically and experimentally that a variety of sizes is necessary to effectively utilize the available vibrator energy. Also, anchor performance (penetration and resulting holding capacity) is dependent upon vibrator power, the supply of energy, the length of shaft, and seafloor properties.

The emplacement of this type of anchor consists of lowering the anchor assembly until it reaches the seafloor. The CEL bottom-resting system is activated upon bottom contact; the OSE system, which does not have a support frame, is activated prior to touchdown. The entire CEL anchor system is considered expendable in water depths greater than 1,000 feet. In lesser depths a second line can be used to retrieve the support frame. The OSE installation technique allows retrieval of the vibrator unit after penetration is complete, because the anchor is lowered with dual

Figure 2.2-1. Deep-water vibrated anchor.

lines with the anchor line being attached to the main shaft below the vibrator.

The CEL anchor has two additional features of interest — remote sensing instrumentation which permits determination of the attitude of the anchor when it rests on the seafloor, and a displacement monitoring system which yields penetration depth and rate.

2.2.2. Advantages and Disadvantages

The principal advantages are: (1) It can accommodate layered seafloors or seafloors with variable resistances, because it has a continuous

---
*Linear accelerators have been designed, but greater success has been achieved with counter-rotating eccentrics.

power application throughout penetration. (2) Penetration rate and amount can be monitored. (3) Confirmation of satisfactory implant is attainable. (4) Holding capacity is reasonably predictable.

The principal disadvantages are: (1) Use is limited to sediments. (2) It is difficult to handle from ship and stabilize on the seafloor. (3) The surface vessel must hold position precisely during penetration to prevent toppling. (4) Operation is limited to seafloors with slopes less then 10°.

2.2.3. History and Status

The present designs function in sediments, attaining moderate holding capacities to water depths of 6,000 feet.

Pile and pipe driving by vibratory means has proven to be feasible on land and in water within the past several years. In 1967 the Ocean Science and Engineering Corporation successfully drove a coring pipe into the seafloor with a vibrator unit in 3,000 feet of water off Madagascar. This, combined with the CEL development of a quick-keying fluke (Smith, 1966) provided the catalyst for beginning work on the vibratory anchor concept. Since then, both surface- and seafloor-powered anchors have been designed and tested.

Recently, MKT Corporation and L. R. Foster Company have introduced hydraulic vibratory pile drivers usable to about 60 feet. However, with minor modifications, a depth of 1,000 feet should be attainable (Schmid, 1969). The feasibility of such a system has been demonstrated by the Institut Francais du Pelrole (IFP) where a "Subsea Vibro-Driver" has been fabricated for use to depths of 650 feet (IFP, 1970). This device is designed to insert a large diameter core tube (12 inches) in sediments. It has been used occasionally to set stake piles for anchors.

2.3. SCREW-IN ANCHORS

2.3.1. Description

A screw-in anchor (augured) is a slender shaft having one or more single-turn helical surfaces. It is, literally, screwed into the soil (see Section 3.14). This type of anchor was originally designed for use on land as a guy anchor for electrical transmission lines. New, suitable equipment has been developed to adapt it for use in the seabed. The primary application is as a pipeline anchor in shallow water. The diameter of the helixes, the number of helixes, the magnitude of downward force applied during penetration, the depth of penetration (by means of modular extensions to the shaft), the applied torque, and the strength of the shaft are varied to adjust to different soil properties.

2.3.2. Advantages and Disadvantages

The principal advantages are: (1) Control of penetration. (2) Monitoring of penetration.

The principal disadvantages are: (1) Limited to use in shallow water. (2) Use is limited to sediments. (3) The surface vessel must hold precise position during installation.

2.3.3. History and Status

Screw-in or augered anchors have only recently been introduced to the ocean environment; however, there was considerable land-based technology. Adaptation for ocean use required only the development of a remote surface-powered driving unit. This anchor type is powered from the surface, and its water depth usage, therefore, depends upon properly transmitting power to the driving unit. The current usable water depth is limited to depths of several hundred feet. The principal uses are at present for pipelines in rather shallow depths (up to about 300 feet) in noncohesive soils.

2.4. DRIVEN ANCHORS

2.4.1. Description

A driven anchor is an anchor that is forced into the seabed by repeated impulsive loads, usually from a hammer. The particular forms are, at present, the stake pile (a single pile), the umbrella pile (a pile with fingerlike flukes that expand umbrella-fashion during the final phase of the driving), and a single-plate anchor that is driven with a mandrel and follower and then keyed by a pull-out load applied through the

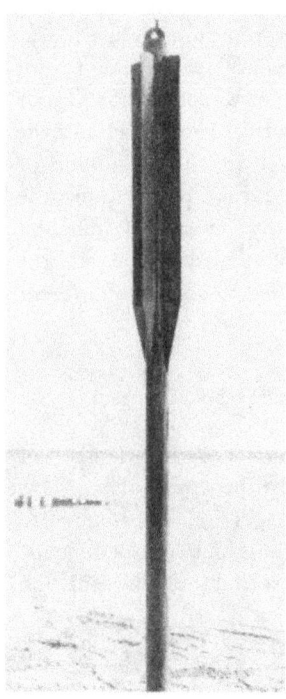

Figure 2.4-1. Driven anchors.

anchor line, as described for the propellant-actuated anchor. Figure 2.4-1 shows a stakepile and one type of umbrella pile. The top of a stake pile (the point of attachment of the anchor line) should be several feet below the seafloor, and the capacity of the stake pile to resist uplift is increased if the load on the pile has a horizontal component. Obviously, flukes minimize this requirement.

2.4.2. Advantages and Disadvantages

The principal advantages are: (1) High capacity in sand. (2) Well established technique. (3) Maximum capacity attained with negligible movement (no keying or setting).

The principal disadvantages are: (1) Limited depth for surface air hammers (about 300 feet). (2) Limited depth for underwater hammers (about 1,000 feet). (3) Requires an enormous amount of surface support. (4) In the case of stake piles, the uplift-resisting capacity is reduced as the resultant load component approaches the vertical.

2.4.3. History and Status

Present technology is limited to rather shallow depths (less than about 300 feet for surface-driven piles and 1,000 feet for underwater driving equipment) because of the present mechanical limitations of hammers and the large mass to be driven.

The state of the art of shallow-water driving is well advanced. Piles, fluked implements, and plates are commonly driven into the seafloor to provide uplift resistance. The driven plate is the most recent usage of the driving technique.

Driving from the surface is the most common and most advanced method of installing piles in the seafloor. Single-acting steam or compressed air hammers and diesel hammers are most often used. The present water depth record for driven piles is 340 feet; plans are underway to extend this record to 1,000 feet in the Santa Barbara Channel.

Subsurface driving is receiving considerable attention because the need for a long follower or expensive templates and surface support is reduced.

Steam or compressed air hammers have been modified for underwater use, and have been utilized for pile driving in a water depth of 163 feet in Narragansett Bay, Rhode Island.

2.5. DRILLED ANCHORS

2.5.1. Description

A drilled anchor is a pile, a length of chain, or other structure that is placed into a previously drilled hole in the seafloor. (See Section 3.14 for an illustration.) Methods for fixing the anchor in the hole include grout (and possibly a technique for expanding the grout against the sides of the hole) and mechanical ears or dogs that are forced outward to engage the sides of the hole when a pull-out load is applied. The technique is intended for rock and coral.

2.5.2. Advantages and Disadvantages

The principal advantage is that it is virtually the only sure type of anchor for rock.

The principal disadvantage is that it requires close control of position during drilling.

2.5.3. History and Status

Drilled and grouted anchors (piles and chains) provide reliable firm anchoring in seafloor rock and soil. Drilling is the only practicable method of emplacing piles in water depths in excess of 600 feet. Actually, drilling has been accomplished to a depth of 12,000 feet. The techniques are basically extensions of offshore oil-drilling methods.

Only a few vessels are available for emplacing pile anchors in very deep water. The Glomar Challenger, the drill ship for the Deep Sea Drilling Project, demonstrated a capability for installing pile anchors at a 20,000-foot depth. Other similar vessels could install piles to 6,000 feet. The major limitation of this anchoring technique is cost, which is up to $15,000/day.

Seafloor rock fasteners (such as rock bolts and grouted rebar), are presently limited to installation by diver and moderate holding capacities. Work to date has been involved with the techniques and equipment to install rock bolts and the shapes of the bolts for various materials. Descriptions and data are included in Section 3.19.

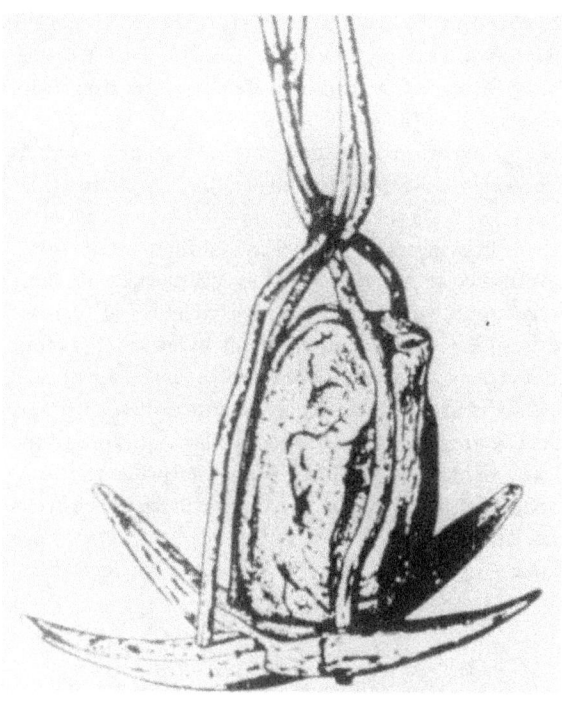

Figure 2.6-1. Primitive deadweight anchor.

2.6. DEADWEIGHT ANCHORS

2.6.1. Description

A deadweight anchor can be any object that is dense, heavy, and resistant to deterioration in water. It is the simplest and most crude form of an anchor. The type of ocean operation and the availability of materials usually dictate the shape, form, size, and weight of a deadweight anchor. Common examples of deadweight anchors are stones, concrete blocks, individual chain links, sections of chain links, and railroad wheels. (Figure 2.6-1 shows a primitive deadweight anchor.) Also, conventional drag-type anchors are sometimes used as deadweight anchors by themselves or in conjunction with other deadweight material.

In most instances a deadweight anchor functions as just that, a deadweight on the seafloor that resists uplift by its own weight in water and resists lateral displacement by its drag coefficient with the seafloor. Deadweight anchors are inefficient and unpredictable. Their drag coefficient varies with the amount of uplift force that coincides with lateral force. On sloping seafloors they tend to slide down slope or are displaced easily when the lateral force component is in the downslope direction. Deadweights are also easily displaced in shallow water by water drag from wave surge.

Conventional anchors are sometimes used as deadweight anchors to combat lateral movement. Of course, this application occurs only in water depths where it is impracticable to embed them by dragging. Conventional anchors used as deadweight anchors resist uplift force by their own weight and increase resistance to lateral displacement by as much as four times over a simple deadweight. A conventional anchor is used effectively in conjunction with simple deadweights by connecting it by chain or cable to the deadweight. The deadweight then provides the resistance to uplift, and the conventional anchor restricts the lateral displacement of the deadweight to a distance no greater than the chain length between them.

In this application a much smaller conventional anchor can be used then if it were used alone.

2.6.2. Advantages and Disadvantages

The principal advantages are: (1) They are simple to construct, economical, and readily available. (2) Their application is independent of most sea-floor conditions, excluding steep, sloping bottoms. (3) Their uplift resistance is precisely predictable. (4) The installation procedures are relatively simple, and the installation equipment required is minimal.

The principal disadvantages are: (1) Their holding-capacity-to-weight ratio is undesirably low. (2) They become increasingly impractical as holding capacity requirements extend beyond 1,500 pounds. (3) They are highly susceptible to unpredictable lateral displacements. (4) They are costly to transport and handle because of their excessive weight.

MAGNAVOX EMBEDMENT ANCHOR, MODEL 1000

Chapter 3. DATA SUMMARIES

This chapter provides data on specific anchor designs that have been developed to meet special needs — primarily, the capability to resist uplift loads; the capability of being rapidly, simply and precisely installed; and the capability of holding in hard material. Such requirements are not satisfied by anchors that must be preset by dragging.

The subsections summarize the data on specific anchor designs. Included are brief comments on the background or the area of use for the anchor, descriptions and details, operational aspects, and cost if established and known. While the details that are pertinent necessarily vary from anchor to anchor, those that are available are fitted into the following outline:

- Source
- General Characteristics
- Details
- Operational Aspects
- Cost
- References

Details on each anchor include such things as advertised nominal holding capacity, nominal penetration, operational depths, advantages and limitations. Care should be exercised in choosing an anchor based upon a company's advertised holding capacity because the capacity may not be necessarily based upon the same assumptions. Actual and estimated holding capacities are summarized in Appendix A and plotted in Appendix B.

Data on "operational modes" are pertinent in the case that more than one method of installation is available, because the method governs such things as speed of installation, precision of placement, and cost. For example, for propellant-actuated anchors, there are two ways to deliver the anchor to the seafloor (referred to as "free-fall" and "cable-lowered"), two ways to activate the firing mechanism ("automatic firing" upon contact with the seafloor and "command-firing" through manual operation of a switch aboard the surface vessel), and two options for dealing with ancillary equipment lowered with the anchor (to recover and reuse it or to abandon it). Of the eight combinations of these procedural options, more than one is often available.

Information not available at this writing, either because it was unknown or could not be obtained within the time frame for this writing, will be indicated by a dash mark. The date of preparation or latest revision is shown at the bottom of each page.

First edition—blanks will be eliminated as revisions are made.   December 1974

MAGNAVOX EMBEDMENT ANCHOR, MODEL 1000

3.1. MAGNAVOX EMBEDMENT ANCHOR SYSTEM, MODEL 1000 (Propellant-Actuated)

3.1.1. Source

The Magnavox Company
1700 Magnavox Way
Fort Wayne, Indiana 46804

3.1.2. General Characteristics

An operational, lightweight, compact, efficient, reliable anchor for use in automatically deployed moorings. It is: (1) suitable for free-fall, unguided, automatic placement, (2) adaptable to systems utilizing manual positioning and remote-command firing, (3) deployable in any depth, and (4) functional in a broad range of seafloors.

Advertised Nominal Holding Capacity

Sandstone and coral:	1,500 lb
Sand:	2,000 lb
Stiff clay:	1,200 lb
Mud and soft clay:	500 lb

Nominal Penetration

Sand:	10 ft
Medium and stiff clays:	6 to 12 ft
Soft silt and clay:	20 ft

Water Depth

Design values —	
Maximum:	20,000 ft
Minimum:	10 ft
Experience —	
Maximum:	13,700 ft
Minimum:	10 ft

Limitations

No known limitations

Advantageous Features

Compact, functional unit

Optional modes of operation, including operation with no line to the surface other than the anchor line

3.1.3. Details

Anchor Assembly (Excluding Lines)

With expendable gun assembly (see Figure 3.1-1) —

Height:	3 ft
Outside diameter:	0.5 ft
Weight:	25 lb

With reusable gun assembly —

Height:	4 ft
Outside diameter:	2.0 ft
Weight:	100 lb

Anchor-Projectile (see Figures 3.1-2 and 3.1-3)

Type:	Streamlined, compact projectile with elongated neck and pointed nose; four outward-opening flukes hinged to the projectile behind the shoulder
Length of projectile:	16 in.
Diameter of projectile behind shoulder:	1.50 in.
Length of fluke:	8 in.
Width of fluke:	1.25 in.
Effective area of flukes:	40 sq in.
Total weight:	3.2 lb

Gun Assembly (see Figure 3.1-2)

Barrel diameter (inside):	0.75 in.
Length of travel:	8.5 in.
Maximum working pressure:	—
Separation velocity:	—
Upward reaction distance:	—
Propellant:	—
Primer:	—

First edition—blanks will be eliminated as revisions are made.

December 1974

MAGNAVOX EMBEDMENT ANCHOR, MODEL 1000

3.1.4. Operational Aspects

Operational Modes

Expendable gun assembly (GA):
1. Free-fall, automatic firing, GA not recovered

Reusable gun assembly (GA):
1. Free-fall, automatic-firing, GA not recovered
2. Free-fall, automatic-firing, GA recovered (primary mode)
3. Cable-lowered, automatic-firing, GA not recovered
4. Cable-lowered, automatic-firing, GA recovered
5. Cable-lowered, command-firing, GA not recovered (unusual)
6. Cable-lowered, command-firing, GA recovered (unusual)

Safety Features

Expendable gun assembly:

Arming wire locks in-line/out-of-line piston in firing mechanism in the "safe" position — extracted just prior to launch

Spring-loaded in-line/out-of-line piston in firing mechanism — aligned when preset hydrostatic pressure is reached

Reusable gun assembly:

Arming wire — as above

Hydrostatic lock — as above

Hydrostatic lock on touchdown probe (Telescoping leg) prevents movement and triggering of firing mechanism

3.1.5. Cost

Number of Anchors	Material Cost per Anchor Installation ($)	
	Reusable Gun Assembly	Expendable Gun Assembly*
5	460	730
10	380	720
100	280	670
500	200	520
1,000	150	370

* Assumes one reusable gun assembly per 100 firings.

3.1.6. References

1. Excerpts from draft copy: Magnavox Self-Embedment Anchor Programs, 1962-1970. Fort Wayne, IN.

2. Letter, C. S. Myers (Magnavox) to R. J. Taylor (CEL), 18 Oct 1973.

3. The Magnavox Company. Brochure FWD539-1: The Magnavox Embedment Anchor System. Fort Wayne, IN., 1974.

MAGNAVOX EMBEDMENT ANCHOR, MODEL 1000

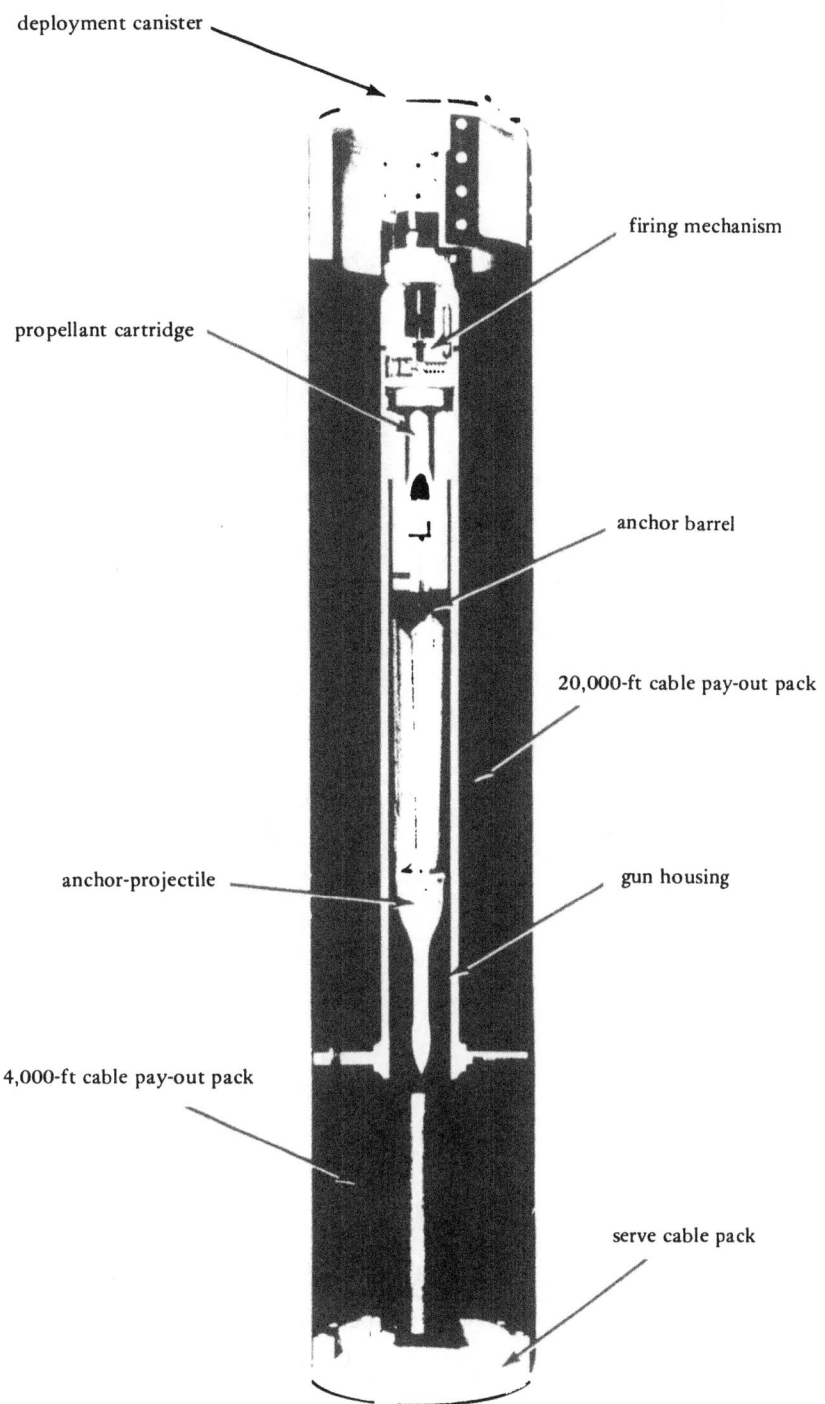

Figure 3.1-1. Magnavox embedment anchor system, Model 1000; cutaway view of anchor-projectile and gun assembly mounted in expendable gun assembly.

December 1974

MAGNAVOX EMBEDMENT ANCHOR, MODEL 1000

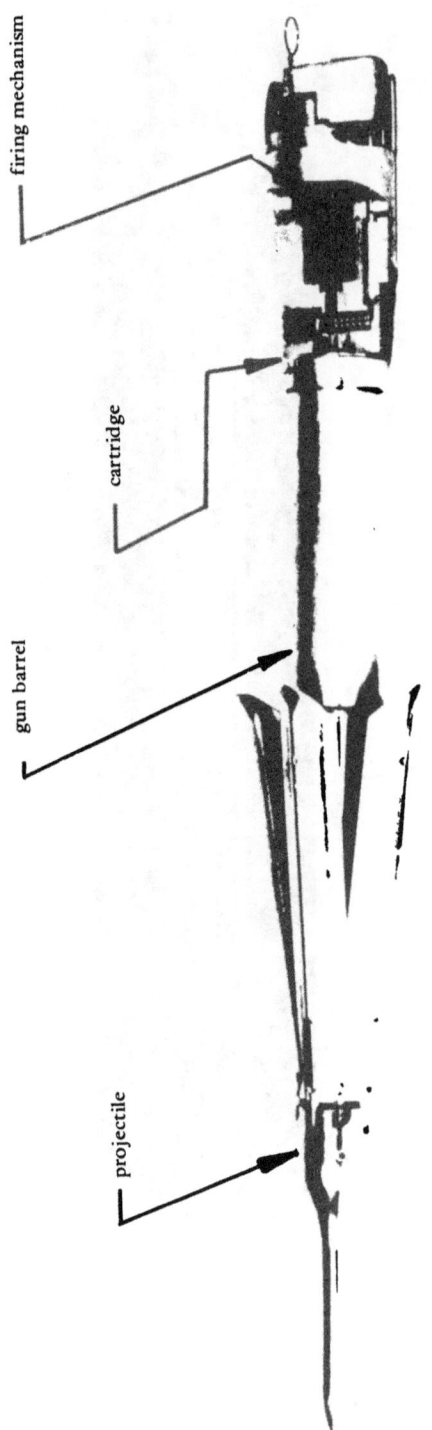

Figure 3.1-2. Magnavox embedment anchor system, Model 1000; without deployment canister.

December 1974

MAGNAVOX EMBEDMENT ANCHOR, MODEL 1000

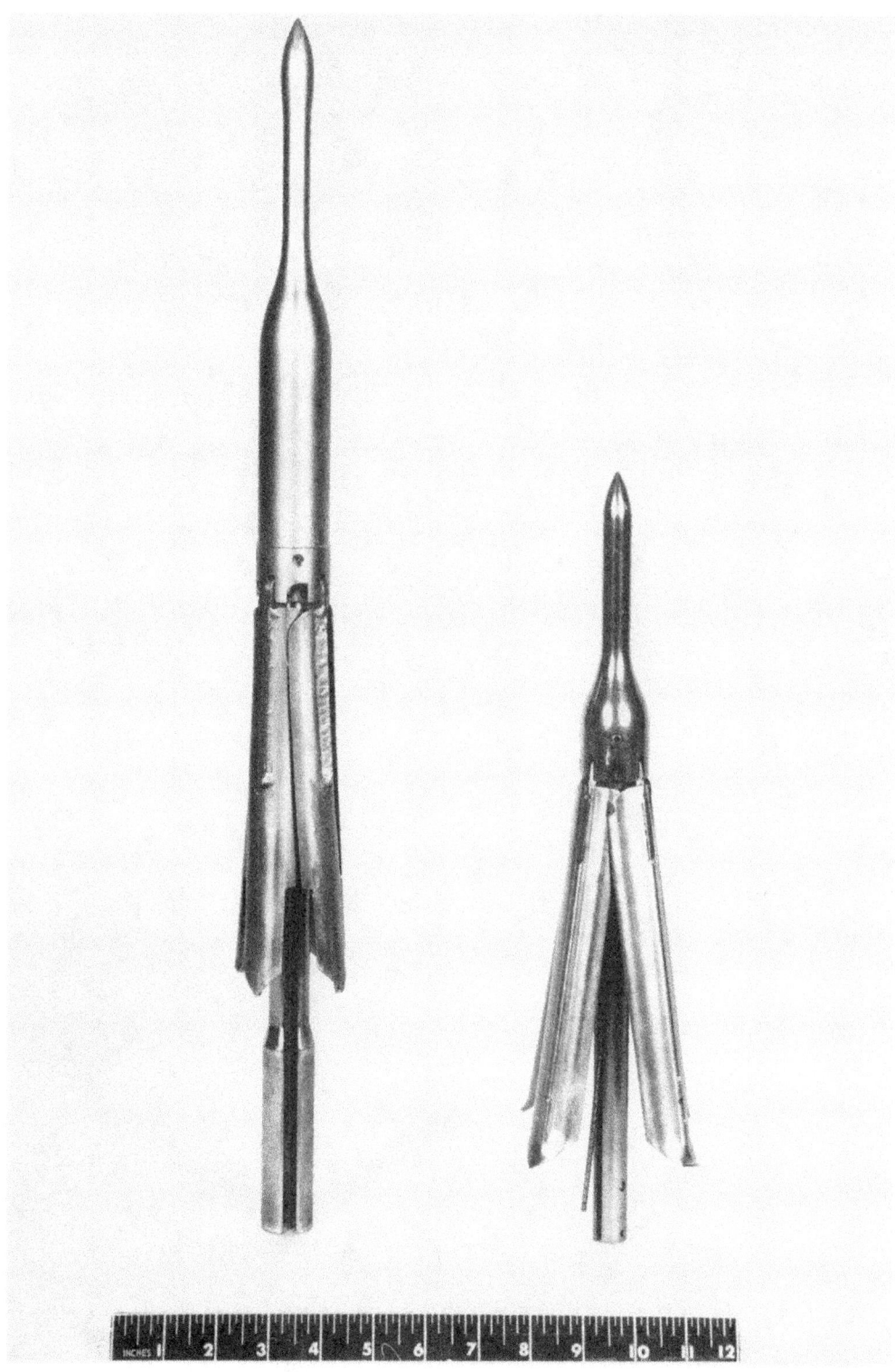

Figure 3.1-3. Magnavox embedment anchor system; anchor-projectiles for Model 1000 (right) and Model 2000 (left).

December 1974

3.2. MAGNAVOX EMBEDMENT ANCHOR SYSTEM, MODEL 2000* (Propellant-Actuated)

3.2.1. Source

The Magnavox Company
1700 Magnavox Way
Fort Wayne, Indiana 46804

3.2.2. General Characteristics

An operational, lightweight, free-fall, propellant-actuated anchor for long-term (3 years) mooring of small navigation buoys in sheltered water with currents of less than 3 knots. It can be: (1) deployed in water 10 feet deep, (2) installed by one man, (3) carried on a 1/2-ton truck and on a small boat, and (4) embedded in a wide range of bottom material.

Advertised Nominal Holding Capacity

Granite:	1,500 lb
Sandstone:	2,000 lb
Coral:	2,000 lb
Sand:	2,000 lb
Stiff clay:	1,700 lb
Mud and soft clay:	800 lb

Nominal Penetration

Silty sand:	10 to 12 ft
Hard clay:	10 to 12 ft
Soft clay and silt:	18 to 20 ft
Very soft silt:	26 to 30 ft

Water Depth

Design values —	
Maximum:	—
Minimum:	10 ft
Experience —	
Maximum:	42 ft
Minimum:	18 ft

Limitations

No known limitations

Advantageous Features

Compact, functional unit

Optional modes of operation, including operation with no line to the surface other than the anchor line

3.2.3. Details

Anchor Assembly (Excluding Lines) (see Figure 3.2-1)

Height:	4 ft
Outside diameter:	2 ft
Weight:	110 lb

Anchor-Projectile (see Figures 3.2-2 and 3.2-3)

Type:	Streamline, compact projectile with elongated neck and bulbous nose with ogibal point; four outward-opening flukes hinged to the projectile behind the shoulder
Length of projectile:	25 in.
Diameter of projectile behind shoulder:	1.5 in.
Length of fluke:	10 in.
Width of fluke:	1.5 in.
Effective area of flukes:	60 sq in.
Total weight:	6.8 lb

Gun Assembly (see Figure 3.2-2)

Barrel diameter (inside):	1.13 in.
Length of travel:	8.1 in.
Maximum working pressure:	60,000 psi
Separation velocity:	500 ft/sec

---

* Data only available for Model 2000, which utilizes a reusable launching system.

First edition—blanks will be eliminated as revisions are made.　　　December 1974

MAGNAVOX EMBEDMENT ANCHOR, MODEL 2000

Upward reaction
    distance: 2.5 ft
Propellant: 500 grains of Hercules HPC 87, 70 grains of Dupont IMR 3031, and 30 grains of Hercules No. 2400
Primer: Federal No. 215

3.2.4. Operational Aspects

Operational Modes

With reusable gun assembly (GA):

1. Free-fall, automatic-firing, GA not recovered
2. Free-fall, automatic-firing, GA recovered (primary mode)
3. Cable-lowered, automatic-firing, GA not recovered
4. Cable-lowered, automatic-firing, GA recovered
5. Cable-lowered, command-firing, GA not recovered
6. Cable-lowered, command-firing, GA recovered (unusual)

Safety Features

Arming wire locks in-line/out-of-line piston in firing mechanism in the "safe" position — extracted just prior to launch

Spring-loaded in-line/out-of-line piston in firing mechanism — aligned when preset hydrostatic pressure is reached

Hydrostatic lock on touchdown probe prevents telescoping and triggering until preset hydrostatic pressure is reached

Shear pin in trigger lever shears if hydrostatic lock does not arm properly

Visual indication of position of in-line/out-of-line piston

3.2.5. Cost

Number of Anchors	Material Cost per Anchor Installation* ($)
5	640
10	500
100	390
500	290
1,000	200

* Assumes one reusable gun assembly per 100 firings.

3.2.6. References

1. The Magnavox Company. Report No. FWD72-115: Explosive embedment anchor development program, by F. L. Erickson. Fort Wayne, IN., Nov 1972. (Contract No. DOT-CG-04468-A)

2. Letter, C. S. Myers (Magnavox) to R. J. Taylor (CEL), 18 Oct 1973.

3. The Magnavox Company. Brochure FWD539-1: The Magnavox Embedment Anchor System. Fort Wayne, IN., 1974.

MAGNAVOX EMBEDMENT ANCHOR, MODEL 2000

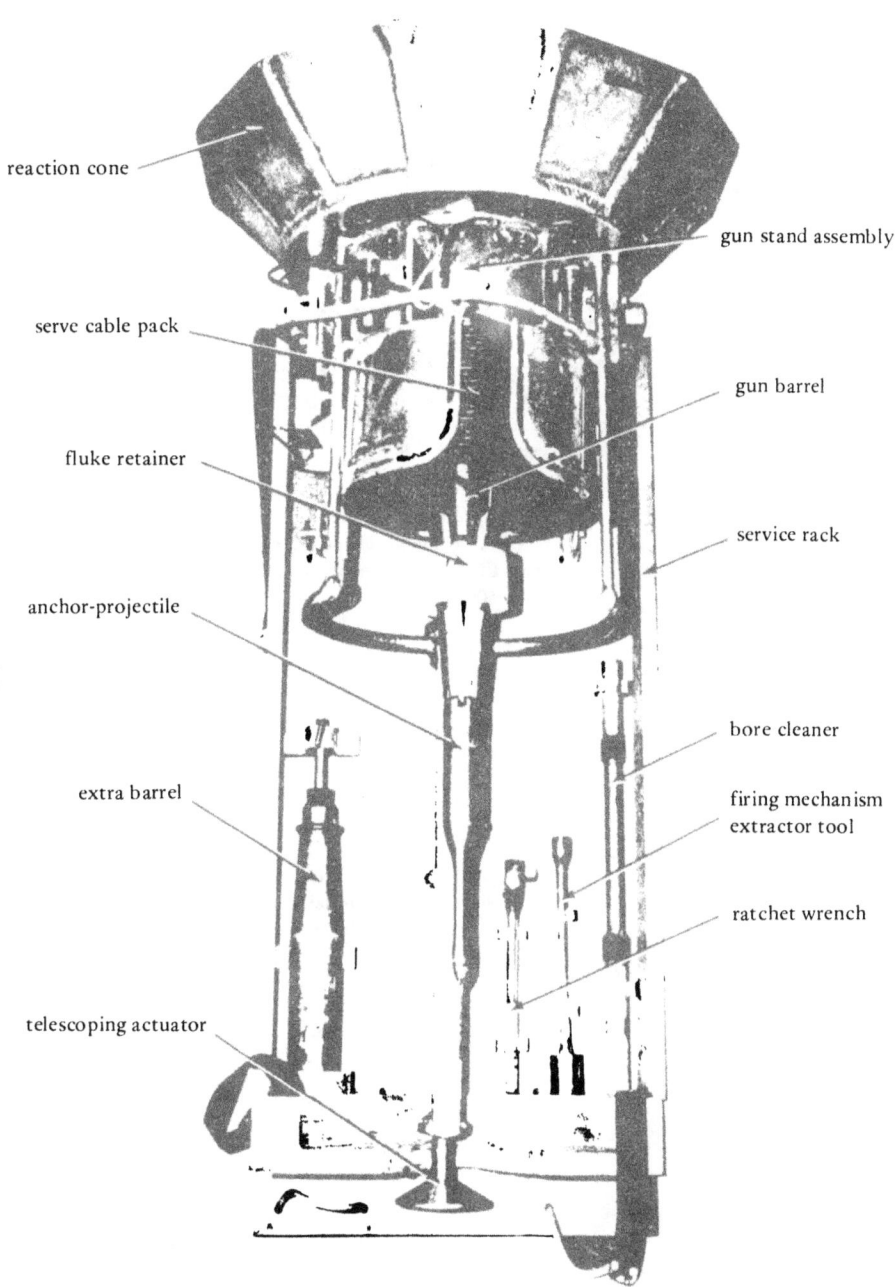

Figure 3.2-1. Magnavox embedment anchor system, Model 2000; reusable gun assembly and accessories.

December 1974

MAGNAVOX EMBEDMENT ANCHOR, MODEL 2000

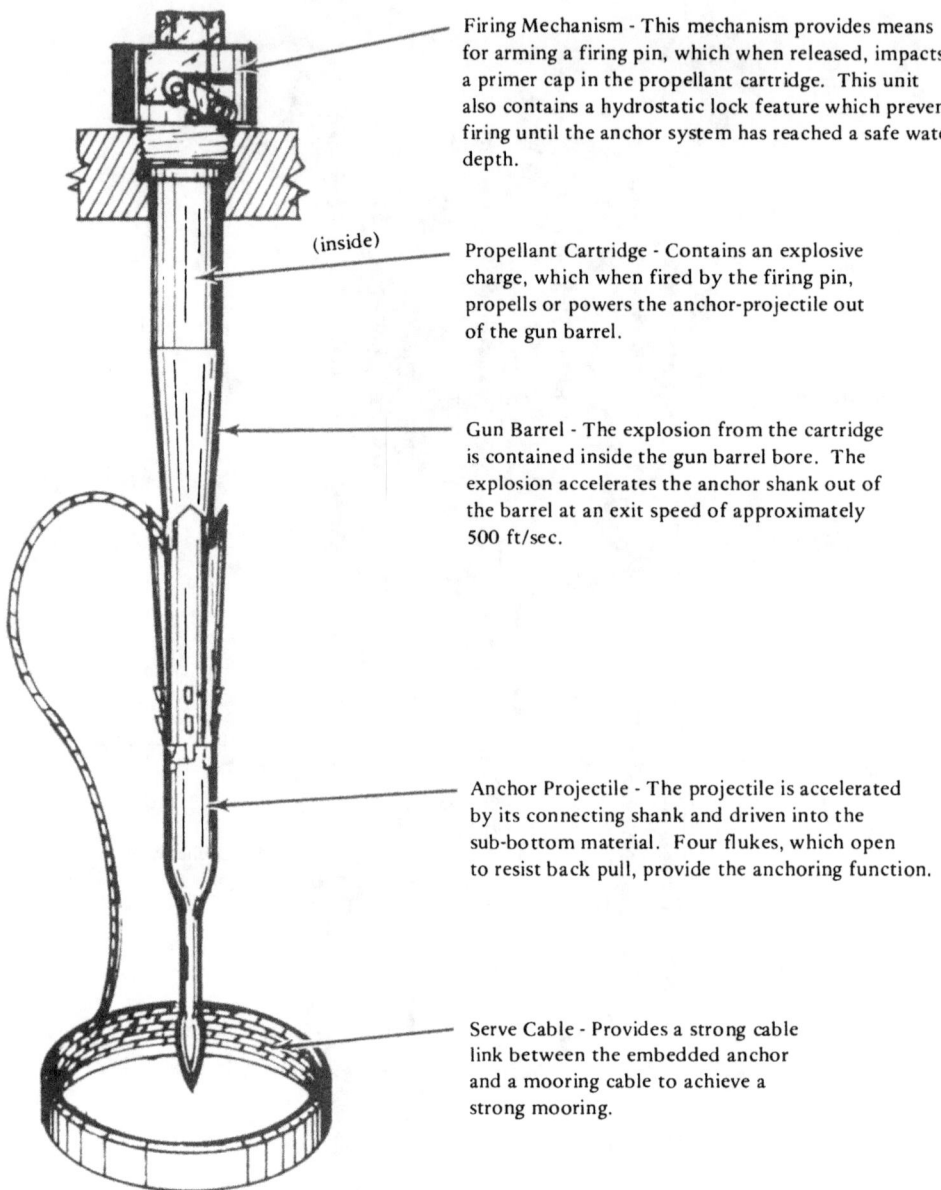

Firing Mechanism - This mechanism provides means for arming a firing pin, which when released, impacts a primer cap in the propellant cartridge. This unit also contains a hydrostatic lock feature which prevents firing until the anchor system has reached a safe water depth.

Propellant Cartridge - Contains an explosive charge, which when fired by the firing pin, propells or powers the anchor-projectile out of the gun barrel.

Gun Barrel - The explosion from the cartridge is contained inside the gun barrel bore. The explosion accelerates the anchor shank out of the barrel at an exit speed of approximately 500 ft/sec.

Anchor Projectile - The projectile is accelerated by its connecting shank and driven into the sub-bottom material. Four flukes, which open to resist back pull, provide the anchoring function.

Serve Cable - Provides a strong cable link between the embedded anchor and a mooring cable to achieve a strong mooring.

Figure 3.2-2. Magnavox embedment anchor system, Model 2000; without reaction cone and gun stand assembly.

December 1974

MAGNAVOX EMBEDMENT ANCHOR, MODEL 2000

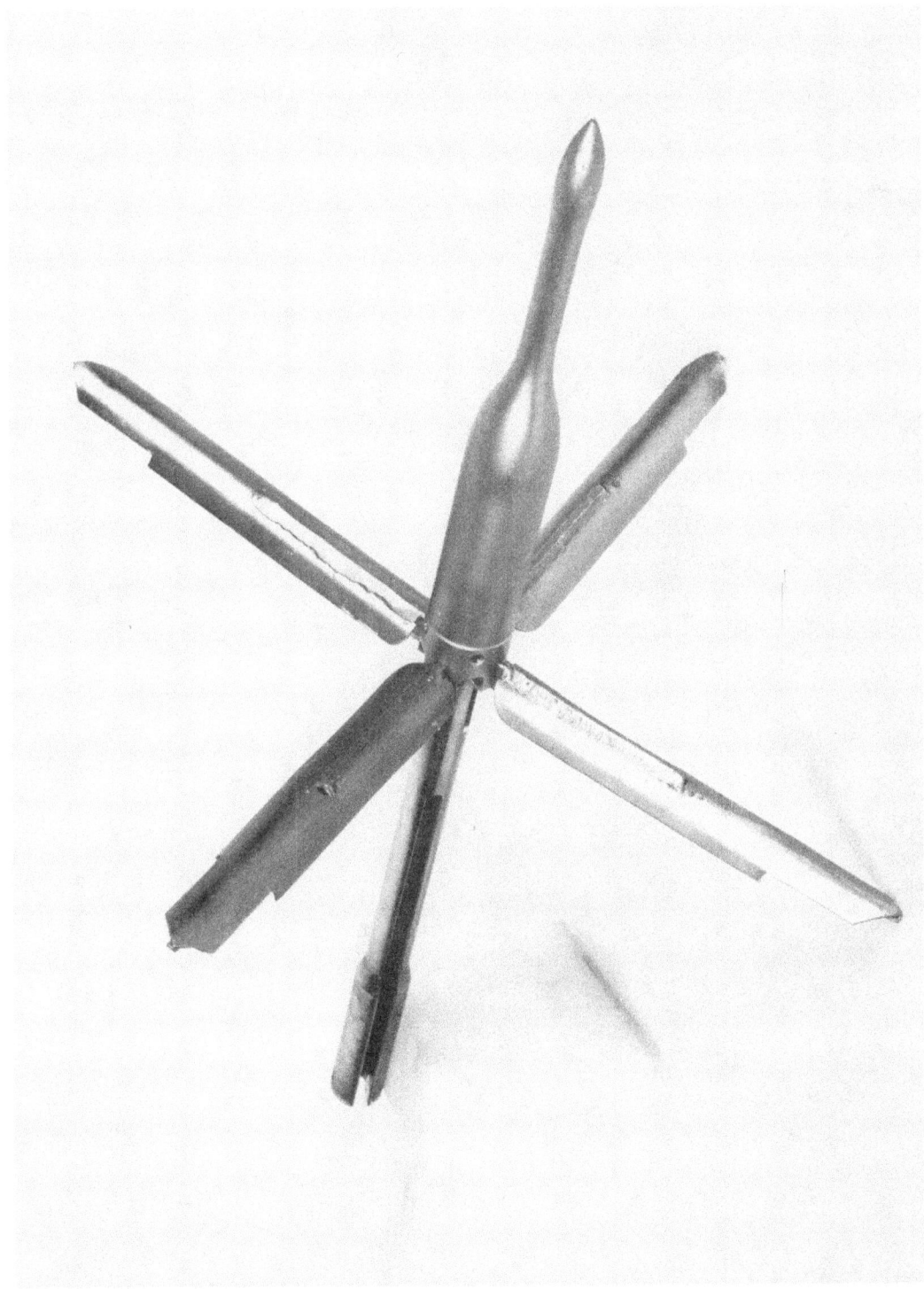

Figure 3.2-3. Magnavox embedment anchor system, Model 2000; anchor with flukes fully opened.

December 1974

VERTOHOLD EMBEDMENT ANCHOR, 10K

3.3. VERTOHOLD EMBEDMENT ANCHOR, 10K
(Propellant-Actuated)

3.3.1. Source

Edo Western Corporation
2645 South 2nd West
Salt Lake City, Utah 84115

3.3.2. General Characteristics

An operational, lightweight anchor for light-to-moderate duty (pipelines, tethered buoys, instruments, pontoon-bridge moorings) and for precise location of anchors in sand, stiff clay, or coral. It has been used for sewer outfalls and buoys.

Advertised Nominal Holding Capacity

10,000 lb

Nominal Penetration

10 to 18 ft

Water Depth

Design values —	
Maximum:	—
Minimum:	—
Experience —	
Maximum:	1,100 ft
Minimum:	45 ft

Limitations

—

Advantageous Features

Optional modes of operation, including operation with no line to the surface other than the anchor line

3.3.3. Details

Anchor Assembly (Excluding Lines) (see Figures 3.3-1 and 3.3-2)

Height:	2.5 ft
Maximum plan dimension (pendant container):	1.5 ft (estimated)
Weight:	60 lb

Anchor-Projectile (see Figures 3.3-1 and 3.3-2)

Type:	Slightly tapered, solid shaft (projectile); two outward-opening flukes hinged to shaft at nose; flukes are flat plates with longitudinal stiffeners; two fluke sizes
Length of projectile:	—
Length of fluke:	14 in.
Width of fluke —	
Anchor for soft material:	5.5 in.
Anchor for hard material:	3.5 in.
Effective area of flukes —	
Anchor for soft material:	154 sq in.
Anchor for hard material:	98 sq in.
Total weight:	25 lb

Gun Assembly

Barrel diameter (inside):	—
Length of travel:	—
Maximum working pressure:	—
Separation velocity:	—
Upward reaction distance:	—
Propellant:	0.34 lb of smokeless powder
Primer:	Shotgun shell

3.3.4. Operational Aspects

Operational Modes

1. Cable-lowered, command-firing, gun assembly recovered (see Figure 3.3-3)
2. Cable-lowered, automatic-firing, gun assembly not recovered (see Figure 3.3-4)

First edition—blanks will be eliminated as revisions are made.

December 1974

VERTOHOLD EMBEDMENT ANCHOR, 10K

Safety Features

    Command-firing mode:

        Safety pin in in-line/out-of-line detonator slide — removed before lowering assembly

        Hydrostatic-pressure actuation of in-line/out-of-line detonator slide

        Shorted-out electrical leads at surface

    Automatic-firing mode:

        Safety pin in detonator slide

        Hydrostatic-pressure actuation of detonator slide

        Safety pin in touchdown probe to prevent movement — removed before lowering assembly

3.3.5. Cost

Number of Anchors	Material Cost per Anchor Installation* ($)
5	775
10	705
25	630
50	560
75	535
100	460

* Gun assembly is recovered; cost of the gun assembly not included.

3.3.6. References

1. Naval Civil Engineering Laboratory. Technical Report R-284-7: Structures in deep ocean; engineering manual for underwater construction, chap 7: Buoys and anchorage systems, by J. E. Smith. Port Hueneme, CA, Oct 1965. (AD473928)

2. Naval Civil Engineering Laboratory. Technical Note N-834: Investigation of embedment anchors for deep ocean use, by J. E. Smith. Port Hueneme, CA, Jul 1966.

3. Naval Civil Engineering Laboratory. Technical Note N-1133: Specialized anchors for the deep sea; progress summary, by J. E. Smith, R. M. Beard, and R. J. Taylor. Port Hueneme, CA, Nov 1970. (AD716408).

4. Naval Civil Engineering Laboratory. Technical Note N-1186: Explosive anchor for salvage operations; progress and status, by J. E. Smith. Port Hueneme, CA, Oct 1971. (AD735104)

5. _____. Technical Note N-1186A: Addendum, by J. E. Smith. Port Hueneme, CA, Jan 1972.

6. Telephone conversation, Mr. Kidd (Edo Western) and Mr. Smith (CEL), 8 May 1972.

7. Edo Western Corporation. Report No. 13076: Operating procedures for Edo Western Corporation's Vertohold embedment anchor. Salt Lake City, UT, Sep 1972.

8. Edo Western Corporation. Pamphlet: Vertohold Embedment Anchors. Salt Lake City, UT, undated.

VERTOHOLD EMBEDMENT ANCHOR, 10K

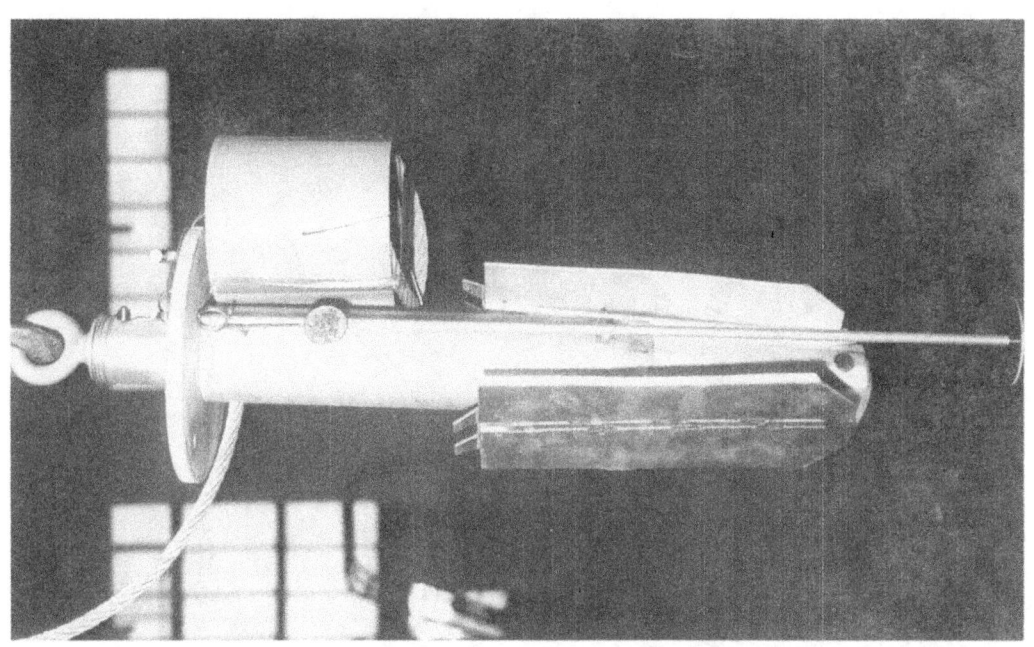

Figure 3.3-2. VERTOHOLD embedment anchor; flukes in penetrating position.

Figure 3.3-1. VERTOHOLD embedment anchor; flukes fully opened.

December 1974

VERTOHOLD EMBEDMENT ANCHOR, 10K

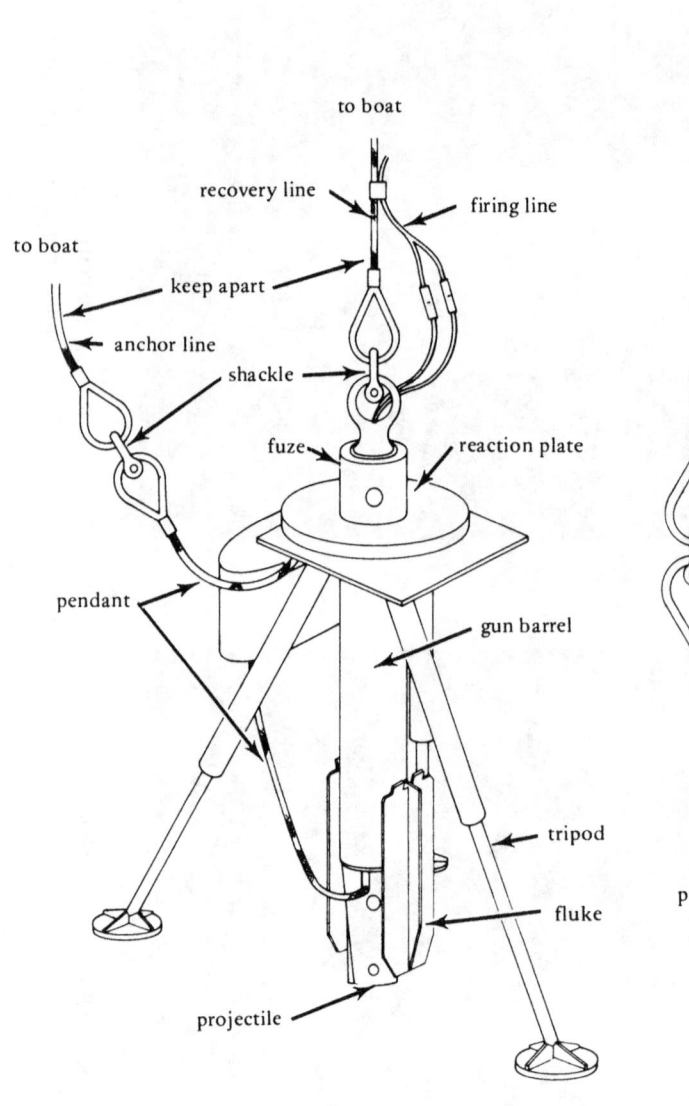

Figure 3.3-3. VERTOHOLD anchor assembly rigged for command-firing and recovery of gun assembly.

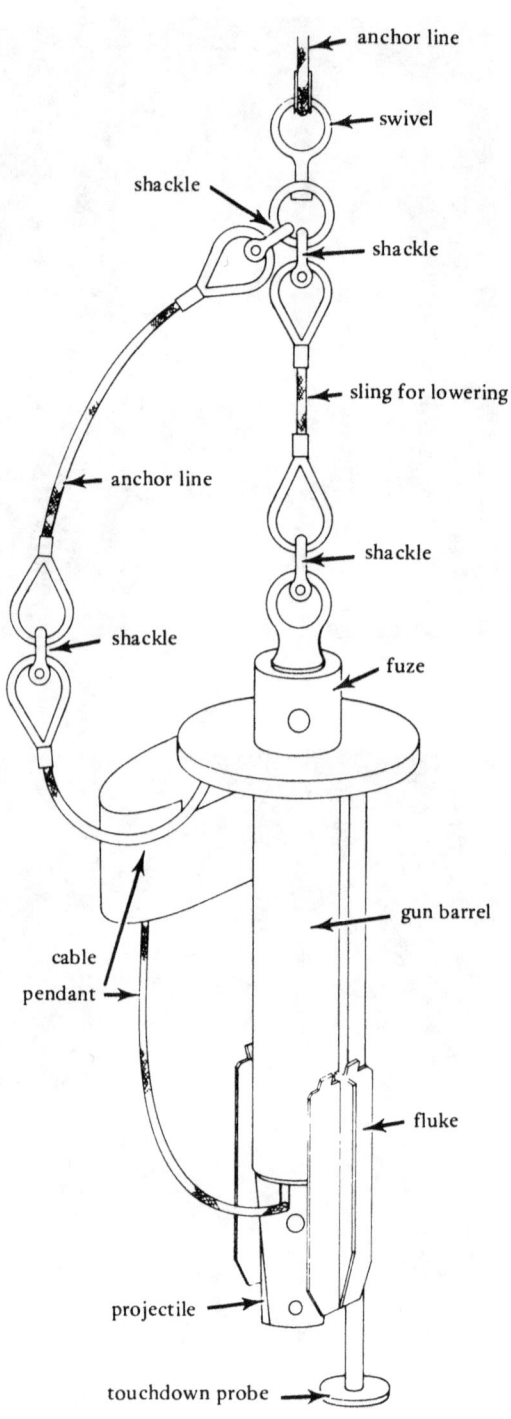

Figure 3.3-4. VERTOHOLD anchor assembly rigged for automatic-firing and nonrecovery of gun assembly.

December 1974

SEASTAPLE EXPLOSIVE EMBEDMENT ANCHOR, MARK 5

3.4. SEASTAPLE EXPLOSIVE EMBEDMENT ANCHOR, MARK 5 (Propellant-Actuated)

3.4.1. Source

Teledyne Movible Offshore, Inc.
P. O. Box 51936 O.C.S.
Lafayette, Louisiana 70501

3.4.2. General Characteristics

An operational, rapidly emplaced, uplift-resisting anchor for precise placement in moderate depths and in any kind of seabed except very hard rock, and for various light-duty applications requiring direct embedment (no dragging for presetting the anchor), such as tiedowns and short-scope moorings.

Advertised Nominal Holding Capacity

5,000 lb

Nominal Penetration

Coral:	2 ft
Sand and medium clay:	7 ft
Mud and soft clay:	20 ft

Water Depth

Design values —	
Maximum:	1,000 ft
Minimum:	10 ft
Experience —	
Maximum:	6,000 ft
Minimum:	10 ft

Limitations

Anchor not usable in rock seafloors

Advantageous Features

Many expensive components are recoverable and reusable (optional)

3.4.3. Details

Anchor Assembly (Excluding Lines) (see Figure 3.4-1)

Height without tripod and probe:	2.25 ft
Height with tripod or probe:	3 ft (estimated)
Reaction cone diameter:	1.1 ft
Maximum plan dimension without tripod (pendant container):	1.5 ft (estimated)
Diameter of tripod foot circle:	6 ft (estimated)
Weight:	60 lb

Anchor-Projectile

Type:	Rotating plate with keying flaps
Length overall:	1.5 ft
Length of fluke:	1.5 ft
Maximum width of fluke:	0.80 ft
Effective area of fluke:	0.83 sq ft
Total weight, including pendant:	10 lb

Gun Assembly

Barrel diameter (inside):	—
Length of travel:	—
Maximum working pressure:	10,000 psi
Separation velocity:	—
Upward reaction distance:	—
Propellant:	0.125 lb
Primer:	—

3.4.4. Operational Aspects

Operational Modes

1. Cable-lowered, automatic-firing, gun assembly not recovered (unusual)
2. Cable-lowered, automatic-firing, gun assembly recovered (primary mode) (see Figure 3.4-2)

First edition—blanks will be eliminated as revisions are made.   December 1974

SEASTAPLE EXPLOSIVE EMBEDMENT ANCHOR, MARK 5

3. Cable-lowered, command-firing, gun assembly not recovered (unusual)
4. Cable-lowered, command-firing, gun assembly recovered

Safety Features

Hydrostatic-pressure actuation of valve in firing mechanism

Safety pin on hydrostatic pressure valve

Shorting and grounding of electrical leads at upper end (command-firing mode)

Safety pin on sliding touchdown probe (automatic-firing mode)

Shielded electrical system

3.4.5. Cost

—

3.4.6. References

1. National Water Lift Company. Operation Instructions: seastaple anchor MK 5-4000. Kalamazoo, MI, Nov 1964.

2. Naval Ordnance Laboratory. Technical Report no. NOLTR 66-205: Field tests to determine the holding powers of explosive embedment anchors in sea bottoms, by J. A. Dohner. White Oak, MD, Oct 1966.

3. Naval Civil Engineering Laboratory. Technical Note N-834: Investigation of embedment anchors for deep ocean use, by J. E. Smith. Port Hueneme, CA, Jul 1966.

4. J. L. Kennedy. "This lightweight explosion-set anchor can stand a big pull," Oil and Gas Journal, vol 67, no. 16, Apr 21, 1969. pp 84-86.

5. "New anchor penetrates rock bottoms," Offshore, vol 29, no. 9, Aug 1969, pp 104, 106-108.

6. Letter, C. D. Ellis (Movible Offshore, Inc.) to J. E. Smith (CEL), Sep 5, 1973.

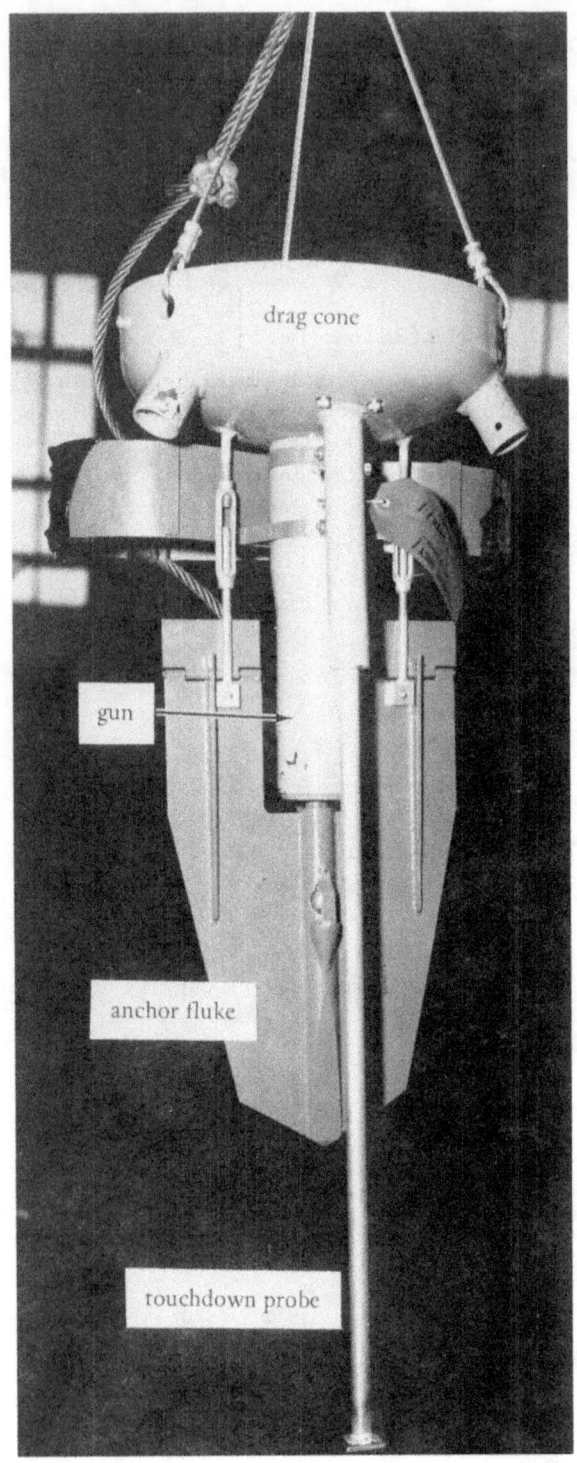

Figure 3.4-1. SEASTAPLE embedment anchor, Mark 5.

First edition—blanks will be eliminated as revisions are made.   December 1974

SEASTAPLE EXPLOSIVE EMBEDMENT ANCHOR, MARK 5

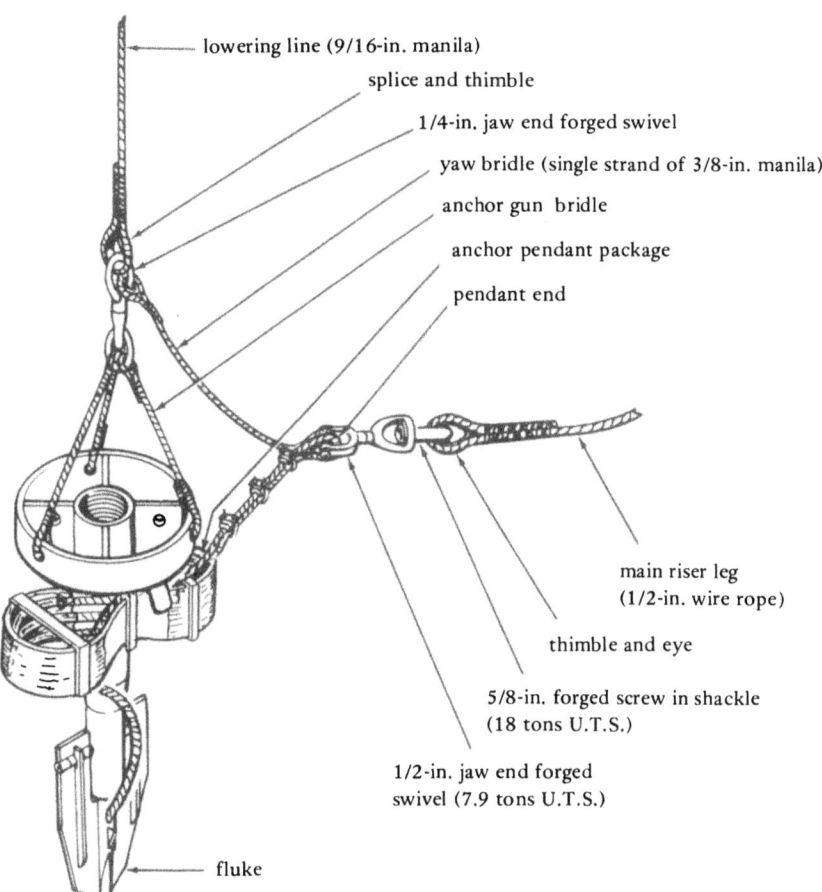

Figure 3.4-2. SEASTAPLE embedment anchor, Mark 5; rigged for recovery of gun assembly.

December 1974

SEASTAPLE EXPLOSIVE EMBEDMENT ANCHOR, MARK 50

3.5. SEASTAPLE EXPLOSIVE EMBEDMENT ANCHOR, MARK 50 (Propellant-Actuated)

3.5.1. Source

Teledyne Movible Offshore, Inc.
P. O. Box 51936 O.C.S.
Lafayette, Louisiana 70501

3.5.2. General Characteristics

An operational, rapidly emplaced uplift-resisting anchor for precise placement in moderate depths and in any kind of seabed except very hard rock, and for short-scope anchoring in various offshore applications (vessels, rigs for offshore oil operations, etc.).

Advertised Nominal Holding Capacity

    50,000 lb

Nominal Penetration

Shale:	4 ft
Sand:	20 ft
Mud:	40 ft

Water Depth

Design values –	
Maximum:	1,000 ft
Minimum:	50 ft
Experience –	
Maximum:	10,000 ft
Minimum:	–

Limitations

    Not usable in competent rock seafloors

Advantageous Features

    Many expensive components are recoverable and reusable (optional)

3.5.3. Details

Anchor Assembly (Excluding Lines) (see Figure 3.5-1)

Height without tripod or probe:	8 ft
Height with tripod or probe:	10 ft (estimated)
Reaction cone diameter:	4 ft
Maximum plan dimension without tripod (pendant container):	4 ft (estimated)
Diameter of tripod foot circle:	18 ft (estimated)
Weight:	1,900 lb

Anchor-Projectile

Type:	Rotating plate with keying flaps
Length overall:	7.5 ft (approx)
Length of fluke:	4.5 ft (approx)
Maximum width of fluke:	2.0 ft (approx)
Effective area of fluke:	8.3 sq ft
Total weight, including pendant:	250 lb

Gun Assembly

Barrel diameter (inside):	5 in.
Length of travel:	38 in.
Maximum working pressure:	–
Separation velocity:	450 ft/sec
Upward reaction distance:	–
Propellant:	3.5 lb
Primer:	–

3.5.4. Operational Aspects

Operational Modes

1. Cable-lowered, automatic-firing, gun assembly not recovered (unusual)
2. Cable-lowered, automatic-firing, gun assembly recovered (primary mode)

First edition–blanks will be eliminated as revisions are made.   December 1974

SEASTAPLE EXPLOSIVE EMBEDMENT ANCHOR, MARK 50

3. Cable-lowered, command-firing, gun assembly not recovered (unusual)
4. Cable-lowered, command-firing, gun assembly recovered

Safety Features

Hydrostatic-pressure actuation of switch valve in firing mechanism

Safety pin on sliding touchdown probe (automatic-firing mode)

Shielded electrical system

3.5.5. Cost

—

3.5.6. References

1. Army Mobility Equipment Research and Development Center. Report no. 1909-A: Development of multi-leg mooring system, Phase A. Explosive embedment anchor, by J. A. Christians and E. D. Meisburger. Fort Belvoir, VA, Dec 1967.

2. J. L. Kennedy. "This lightweight explosion-set anchor can stand a big pull," Oil and Gas Journal, vol 67, no. 16, Apr 21, 1969, pp 84-86.

3. "New anchor penetrates rock bottoms," Offshore, vol 29, no. 9, Aug 1969, pp 104, 106-108.

4. Letter, C. D. Ellis (Movible Offshore, Inc.) to J. E. Smith (CEL), Sep 5, 1973.

SEASTAPLE EXPLOSIVE EMBEDMENT ANCHOR, MARK 50

Figure 3.5-1. SEASTAPLE embedment anchor, Mark 50.

CEL 20K PROPELLANT ANCHOR

3.6. CEL 20K PROPELLANT ANCHOR
(Propellant-Actuated)

3.6.1. Source

Civil Engineering Laboratory
Naval Construction Battalion Center
Port Hueneme, California 93043

3.6.2. General Characteristics

An operational, direct-embedment anchor system of minimum cost for use in very deep water in short-scope moorings and other applications requiring significant resistance to uplift (see Figure 3.6-1).

Advertised Nominal Holding Capacity

20,000 lb

Nominal Penetration

Basalt:	2 ft
Sand:	20 ft
Medium clay:	40 ft

Water Depth

Design values —	
Maximum:	20,000 ft
Minimum:	90 ft
Experience —	
Maximum:	18,600 ft
Minimum:	50 ft

Limitations

No known limitations

Advantageous Features

The system, which is inexpensive to fabricate, is expendable in deep water. Surplus Army or Navy gun barrels are used.

3.6.3. Details

Anchor Assembly (Excluding Lines) (see Figures 3.6-1 and 3.6-2)

Height without touchdown probe:	7 ft*
Height with touchdown probe:	9 ft*
Maximum plan dimension without cable-mounting board:	2.5 ft
Maximum plan dimension with cable-mounting board	3.5 ft
Weight:	1,800 lb**

* Add 2 ft for mud fluke.
** Add 200 lb for mud fluke.

Anchor-Projectile

For rock and coral (see Figure 3.6-3) —

Type:	Round shaft with tapered nose (1:6) and three tapered fins (1:2) Primary fins spaced at 140°
Length of projectile:	3 ft
Length of fins:	2.5 ft
Diameter of shaft:	3 in.
Diameter of circumscribing cylinder:	27 in.
Thickness of fins:	1 in.
Weight, including piston (115 lb):	275 lb

For sand and stiff clay (see Figures 3.6-4 and 3.6-5) —

Type:	Rotating plate
Length overall:	38 in.
Length of fluke:	38 in.
Width of fluke:	18 in.
Effective area of fluke:	4.5 sq ft
Total weight, including piston:	300 lb

First edition—blanks will be eliminated as revisions are made.

December 1974

CEL 20K PROPELLANT ANCHOR

For sand and clay (2 x 4 ft) —
Type:	Rotating plate
Length overall:	51 in.
Length of fluke:	51 in.
Width of fluke:	24 in.
Effective area of fluke:	8.0 sq ft
Total weight, including piston:	370 lb

For mud and soft clay (2-1/2 x 5 ft) —
Type:	Rotating plate
Length overall:	63 in.
Length of fluke:	63 in.
Width of fluke:	30 in.
Effective area of fluke:	12.5 sq ft
Total weight, including piston:	490 lb

Gun Assembly
Barrel diameter (inside):	4.25 in.
Length of travel:	26 in.
Maximum working pressure:	35,000 psi
Separation velocity:	400 ft/sec
Upward reaction distance:	25 ft
Propellant:	3.75 lb max of Standard Navy pyrotechnic (smokeless)
Primer:	M-58 (black powder)

3.6.4. Operational Aspects

Operational Modes

1. Cable-lowered, automatic-firing, gun assembly not recovered (see Figure 3.6-1)
2. Cable-lowered, automatic-firing, gun assembly recovered

Safety Features

Safety pin holds in-line/out-of-line plunger in safe-and-arm device out of line — extracted prior to lowering

Hydrostatic-pressure actuation of in-line/out-of-line plunger

Hydrostatic-pressure actuation of switch in power package

3.6.5. Cost

The material cost per anchor installation when purchased in lots of from 1 to 20 anchors is:

$1,360	With reusable gun assembly (assumes one reusable gun assembly per 20 firings; the gun assembly cost is $3,200)
$4,500	With expendable gun assembly

3.6.6. References

1. Naval Civil Engineering Laboratory. Technical Note N-1282: Propellant-actuated deep water anchor; interim report, by R. J. Taylor and R. M. Beard. Port Hueneme, CA, Aug 1973. (AD765570)

First edition—blanks will be eliminated as revisions are made.   December 1974

CEL 20K PROPELLANT ANCHOR

Figure 3.6-1. CEL 20K Propellant Anchor.

December 1974

CEL 20K PROPELLANT ANCHOR

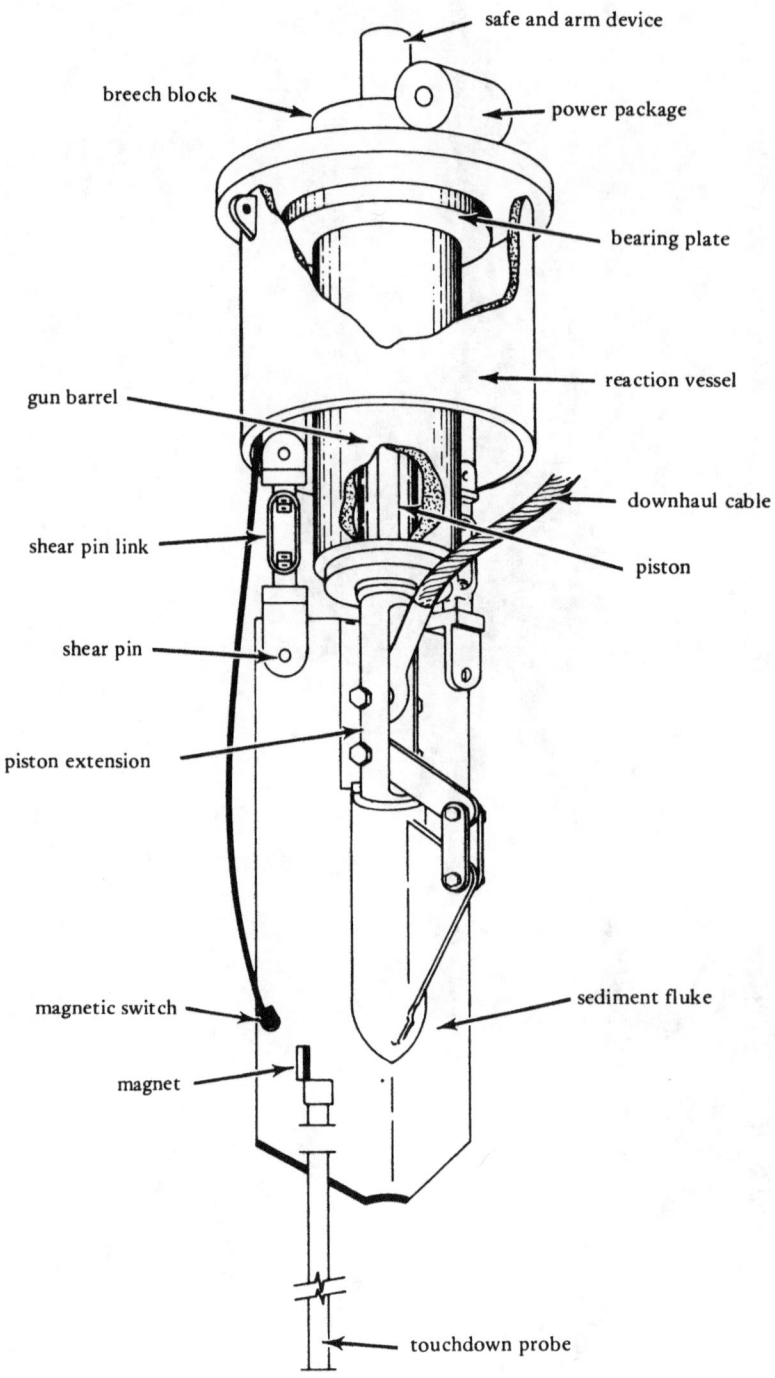

Figure 3.6-2. Schematic of CEL 20K Propellant Anchor.

December 1974

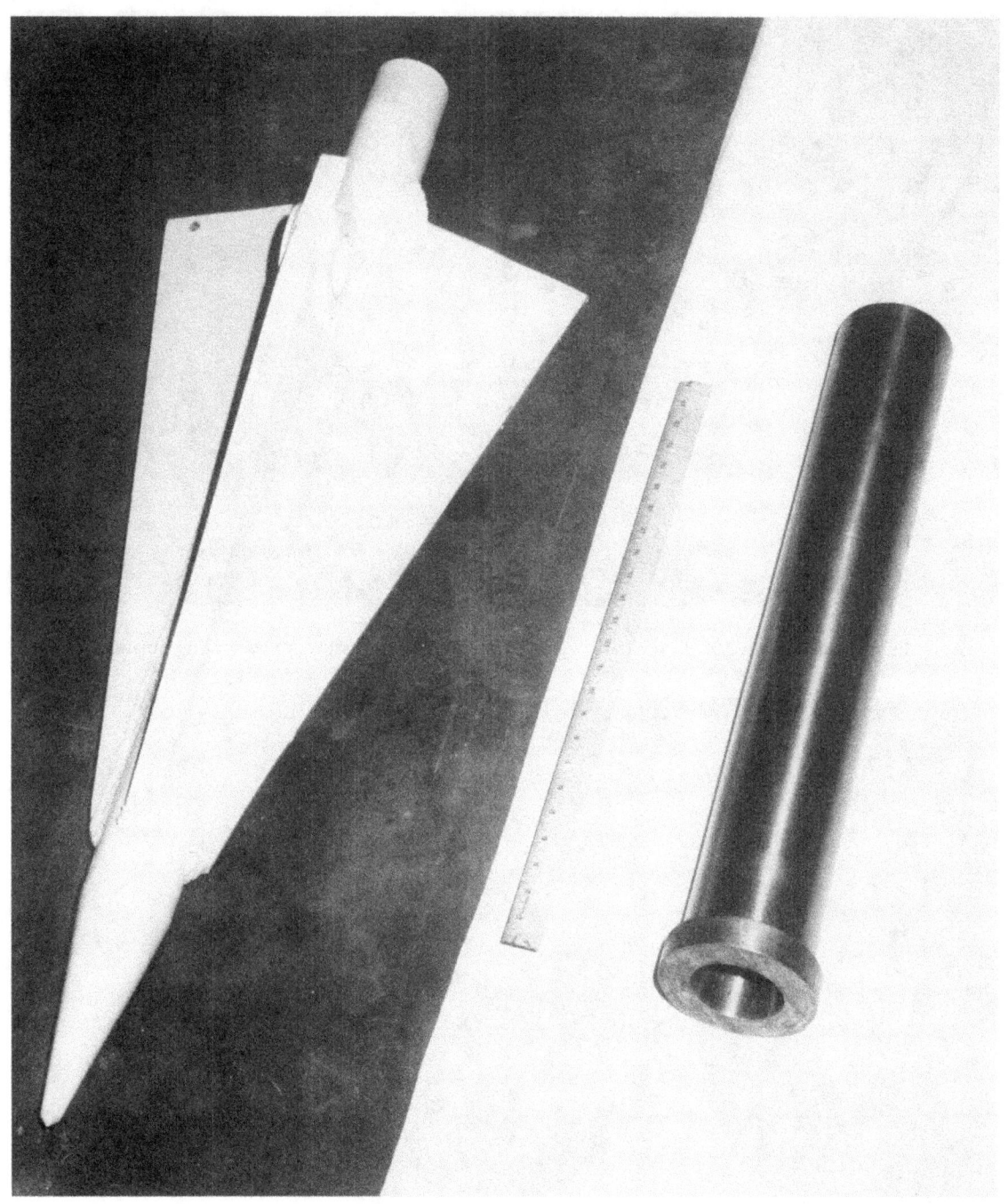

Figure 3.6-3. CEL 20K Propellant Anchor; rock fluke and piston.

CEL 20K PROPELLANT ANCHOR

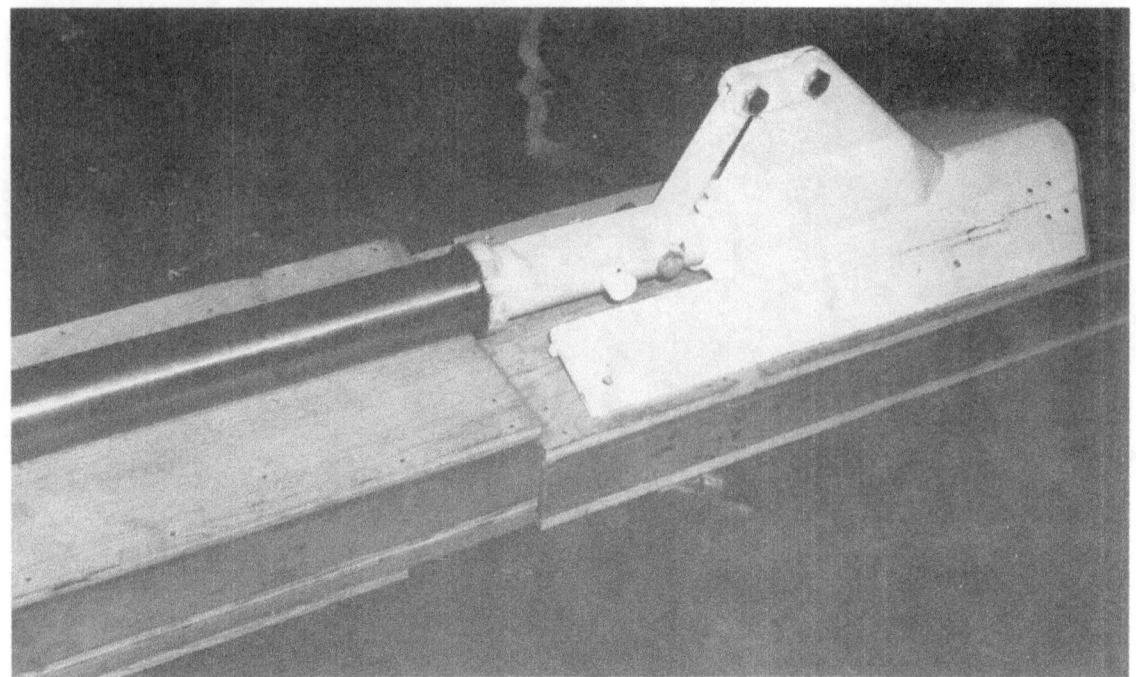

Figure 3.6-4. CEL 20K Propellant Anchor; sand fluke and piston in penetrating position.

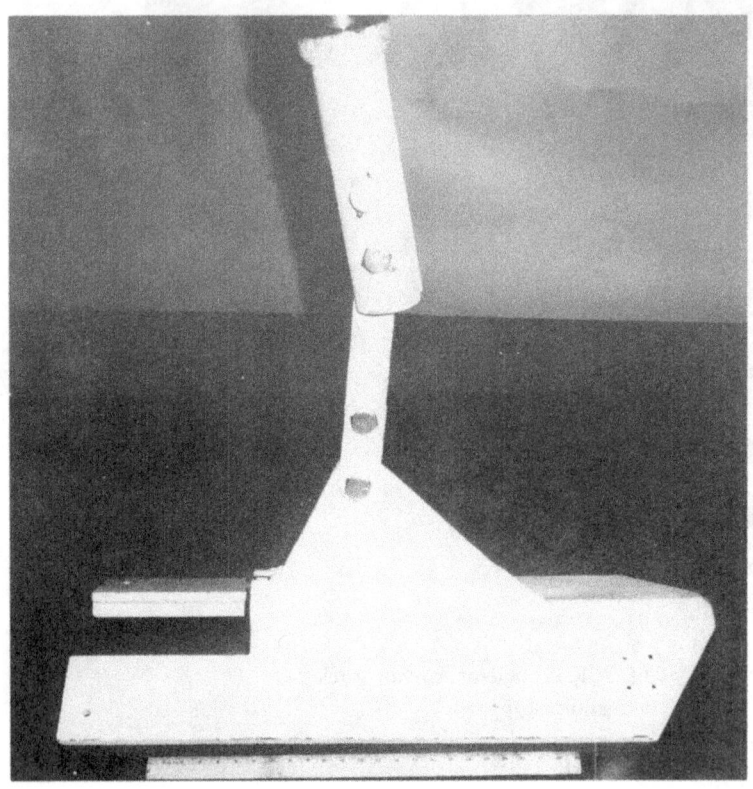

Figure 3.6-5. CEL 20K Propellant Anchor; sand fluke and piston in keyed position.

CEL 100K PROPELLANT ANCHOR

3.7. CEL 100K PROPELLANT ANCHOR
(Propellant-Actuated)

3.7.1. Source

Civil Engineering Laboratory
Naval Construction Battalion Center
Port Hueneme, California 93043

3.7.2. General Characteristics

An operational anchor that is undergoing further testing of a revised launch vehicle design and a new sediment anchor-projectile design. The anchor, which is for use in ship-salvage operations, (1) can be placed rapidly without dragging, (2) develops the full working strength of a standard Navy beach gear leg, and (3) can be handled on an ARS, ASR, ATF, or ATS.

Advertised Nominal Holding Capacity

100,000 lb

Nominal Penetration

Vesicular basalt:	2 ft
Coral:	7 ft
Sand:	20 to 30 ft
Mud:	30 to 50 ft

Water Depth

Design values –
Maximum:	500 ft
Minimum:	50 ft

Experience –
Maximum:	700 ft
Minimum:	35 ft

Limitations

Seafloor must be level and smooth enough to assure upright attitude of launch vehicle (tilt less than 30 degrees)

Potential entanglement problems with multiple lines

Advantageous Features

Stable, rugged launch vehicle

Total manual control of placement and firing, which permits interruptions to assure correctness of operation

High capacity in sand, coral, and rock

Many expensive components are recoverable and reusable

3.7.3. Details

Anchor Assembly (Excluding Lines) (see Figure 3.7-1)

Height:	11 ft
Plan dimension:	8 sq ft
Weight:	13,000 lb

Anchor-Projectile

For rock and coral –

Type:	Three fixed fins (Y-section with 120-degree dihedral angles); fins tapered
Length of projectile:	6.75 ft
Length of fins:	5.0 ft
Diameter of circumscribing cylinder:	3.1 ft
Weight, including piston (500 lb):	2,000 lb

For sand and coral (new design) (see Figure 3.7-2) –

Type:	Rotating plate
Length of projectile:	5-1/2 ft
Length of fluke:	5-1/2 ft
Width of fluke:	2-3/4 ft
Effective area of fluke:	13 sq ft
Weight, including piston (500 lb):	1,550 lb

For clay (new design) (see Figure 3.7-2) –

Type:	Rotating plate
Length of projectile:	6-2/3 ft
Length of fluke:	6-2/3 ft
Width of fluke:	3-1/3 ft

First edition—blanks will be eliminated as revisions are made.

December 1974

CEL 100K PROPELLANT ANCHOR

Effective area of
fluke: 22 sq ft
Weight, including
piston (500 lb): 1,900 lb

Gun Assembly

Barrel diameter
(inside): 10 in.
Length of travel: 36 in.
Maximum working
pressure: 35,000 psi
Separation velocity: 350 to 400 ft/sec
Upward reaction
distance: 8 to 12 ft
Propellant: 14 lb of M26 smokeless powder
Primer: M58 (black powder)

3.7.4. Operational Aspects

Operational Modes

1. Cable-lowered, command-firing, launch vehicle recovered

Safety Features

Lanyard-operated safety pin – pulled as launch vehicle leaves the deck

Interrupted explosive train, with in-line/out-of-line plunger controlled by switch aboard ship and by hydrostatic pressure

Visual safe-arm indication

3.7.5. Cost

The material cost per anchor installation when purchased in lots of from 1 to 20 anchors is $4,100. This assumes one reusable gun assembly per 20 firings with the gun assembly cost being $17,000. The piston, which costs $1,500, is recovered 50% of the time. This cost also includes $1,000 for an expendable anchor pendant.

3.7.6. References

1. Naval Ship Systems Command. Supervisor of Salvage. NAVSHIPS 0994-007-1010: Technical manual: Assembly, stowage, and operation; Anchor, salvage embedment. Washington, DC, Jan 1970.

2. Naval Civil Engineering Laboratory. Technical Note N-1186: Explosive anchor for salvage operations; progress and status, by J. E. Smith. Port Hueneme, CA, Oct 1971. (AD735104)

3. ———. Technical Note N-1186A: Addendum, by J. E. Smith. Port Hueneme, CA, Jan 1972.

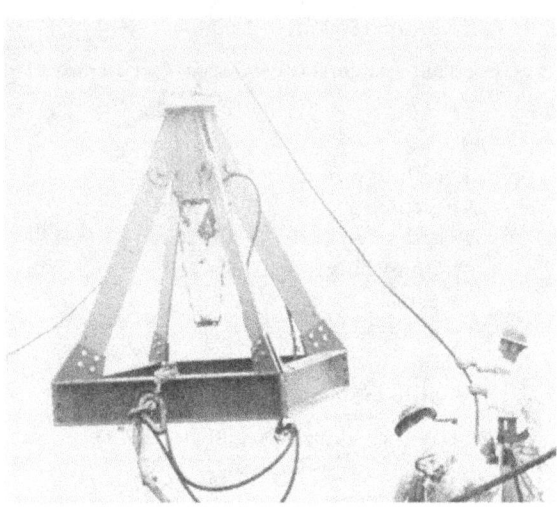

Figure 3.7-1. CEL 100K Propellant Anchor; launch vehicle with dummy anchor (used to evaluate gun performance).

Figure 3.7-2. CEL 100K Propellant Anchor; flukes.

First edition—blanks will be eliminated as revisions are made.

December 1974

EXPLOSIVE EMBEDMENT ANCHOR, XM-50

3.8. EXPLOSIVE EMBEDMENT ANCHOR, XM-50
(Propellant-Actuated)

3.8.1. Source

U.S. Army Mobility Equipment Research and
Development Center
Code SMEFB-HP
Fort Belvoir, Virginia 22060

3.8.2. General Characteristics

An operational, lightweight anchor (relative to a conventional drag-type anchor of comparable capacity) that uses rope instead of chain in multileg moorings for 25,000-DWT tankers in shallow, exposed coastal waters (maximum wave height, 11 ft). It is (1) reliable, (2) quickly installed by Army personnel, (3) suitable for any kind of seafloor material except consolidated rock, (4) adaptable to fleet-type single-point moorings, (5) and air-transportable (C-130).

Advertised Nominal Holding Capacity

50,000 lb

Nominal Penetration

Coral:	20 ft
Sand:	20 ft
Mud and soft clay:	40 ft

Water Depth

Design values —	
Maximum:	150 ft
Minimum:	25 ft
Experience —	
Maximum:	51 ft
Minimum:	9 ft

Limitations

Not usable in competent rock seafloors

Advantageous Features

Many expensive components are recoverable and reusable

3.8.3. Details

Anchor Assembly (Excluding Lines) (see Figures 3.8-1 and 3.8-2)

Height, including probe extension (2.5 ft):	12.2 ft
Drag cone diameter:	4.0 ft
Weight, including riser cable (70 lb):	1,900 lb

Anchor-Projectile (see Figures 3.8-1 and 3.8-2)

Type:	Rotating plate
Length overall:	4.83 ft
Length of fluke:	4.83 ft
Width of fluke:	2.0 ft
Effective area of fluke:	8 sq ft
Weight (includes piston):	400 lb

Gun Assembly

Barrel diameter (inside):	5 in.
Length of travel:	38 in.
Maximum working pressure:	53,000 psi
Separation velocity:	400 to 500 ft/sec
Upward reaction distance:	10 ft
Propellant:	3.5 lb of M2 (MIL-P-323)
Primer:	Two WOX69A (Navy Mk 101) and 6 to 7 ft of Dupont Pyrocore no. 2040 cord

3.8.4. Operational Aspects

Operational Modes

Cable-lowered, automatic-firing, gun assembly recovered

First edition—blanks will be eliminated as revisions are made. December 1974

EXPLOSIVE EMBEDMENT ANCHOR, XM-50

Safety Features

Hydrostatic-pressure actuation of switch in fuze

3.8.5. Cost

The material cost per anchor installation when purchased in lots of from 1 to 20 anchors is $4,750. This assumes one reusable gun assembly per 20 firings with the gun assembly costing $3,000. The polyurethane-coated nylon pendant, costing $1,000, is expendable.

3.8.6. References

1. Army Mobility Equipment Research and Development Center. Report No. 1909-A: Development of multi-leg mooring system, Phase A. Explosive embedment anchor, by J. A. Christians and E. P. Meisburger. Fort Belvoir, VA, Dec 1967.

2. Letter, Commander U.S. Army CDC to Distribution H, 1 Nov 1972, subject: Revised Department of the Army Approved Qualitative Material Requirement (QMR) for multi-leg tanker mooring system.

3. H. C. Mayo. "Explosive anchors for ship mooring," Marine Technology Society Journal, vol 7, no. 6, Sep 1973, pp 27-34.

4. Army Mobility Equipment Research and Development Center. Report No. 2078: Explosive embedment anchors for ship mooring, by H. C. Mayo. Fort Belvoir, VA, Nov 1973.

5. Letter, Commander U.S. Army MERDC to Commander NCEL, 25 Feb 1974, subject: MERDC explosive anchor.

Figure 3.8-1. Army explosive embedment anchor, XM-50; front quarter view.

Figure 3.8-2. Army explosive embedment anchor, XM-50; rear quarter view.

EXPLOSIVE EMBEDMENT ANCHOR, XM-200

3.9. EXPLOSIVE EMBEDMENT ANCHOR, XM-200
(Propellant-Actuated)

3.9.1. Source

U.S. Army Mobility Equipment Research and Development Center
Code SMEFB-HP
Fort Belvoir, Virginia 22060

3.9.2. General Characteristics

An operational anchor that uses rope instead of chain in multileg moorings for 40,000-DWT tankers in shallow, sheltered coastal waters (maximum wave height, 3 ft). The anchor is (1) reliable, (2) quickly installed by Army personnel, (3) suitable for any kind of seafloor material except consolidated rock, (4) adaptable for fleet-type single-point moorings, and (5) a component of a mooring system that is a significantly smaller logistic burden than systems using conventional drag-type anchors.

Advertised Nominal Holding Capacity

200,000 lb

Nominal Penetration

Coral:	15 ft
Sand and stiff clay:	20 ft
Mud and soft clay:	30 to 40 ft

Water Depth

Design values —	
Maximum:	150 ft
Minimum:	40 ft
Experience —	
Maximum:	55 ft
Minimum:	36 ft

Limitations

Not usable in competent rock seafloors

Advantageous Features

Many expensive components are recoverable and reusable

3.9.3. Details

Anchor Assembly (Excluding Lines) (see Figure 3.9-1)

Height, including probe extension (4.0 ft):	18.0 ft
Drag cone diameter:	4.0 ft
Weight, including riser cable (1,200 lb):	5,300 lb

Anchor-Projectile (see Figures 3.9-1 and 3.9-2)

Type:	Rotating plate
Length overall:	6.6 ft
Length of fluke:	6.6 ft
Width of fluke:	3.5 ft
Effective area of fluke:	20 sq ft
Weight (including piston):	1,200 lb

Gun Assembly (see Figure 3.9-3)

Barrel diameter (inside):	6 in.
Length of travel:	60 in.
Maximum working pressure:	60,000 psi
Separation velocity:	400 ft/sec
Upward reaction distance:	30 ft
Propellant:	14 lb of Pyrocellulose (Navy 8/55 smokeless)
Primer:	Two WOX69A (Navy Mk 101) and 9 ft of Dupont Pyrocore No. 2040 cord

3.9.4. Operational Aspects

Operational Modes

Cable-lowered, automatic-firing, gun assembly recovered

First edition—blanks will be eliminated as revisions are made.

December 1974

EXPLOSIVE EMBEDMENT ANCHOR, XM-200

Safety Features

Hydrostatic-pressure actuation of switch in fuze

3.9.5. Cost

The material cost per anchor installation when purchased in lots of from 1 to 20 anchors is $11,450. This assumes one reusable gun assembly per 20 firings with the gun assembly costing $9,000. The polyurethane-coated nylon pendant, costing $1,000, is expendable.

3.9.6. References

1. Army Mobility Equipment Research and Development Center. Report No. 1909-A: Development of multi-leg mooring system, Phase A. Explosive embedment anchor, by J. A. Christians and E. P. Meisburger. Fort Belvoir, VA, Dec 1967.

2. Letter, Commander U.S. Army CDC to Distribution H, 1 Nov 1972, subject: Revised Department of the Army approved Qualitative Material Requirement (QMR) for multi-leg tanker mooring system.

3. H. C. Mayo. "Explosive anchors for ship mooring," Marine Technology Society Journal, vol 7, no. 6, Sep 1973, pp 27-34.

4. Army Mobility Equipment Research and Development Center. Report No. 2078: Explosive embedment anchors for ship mooring, by H. C. Mayo. Fort Belvoir, VA, Nov 1973.

5. Letter, Commander U.S. Army MERDC to Commander NCEL, 25 Feb 1974, subject: MERDC explosive anchor.

EXPLOSIVE EMBEDMENT ANCHOR, XM-200

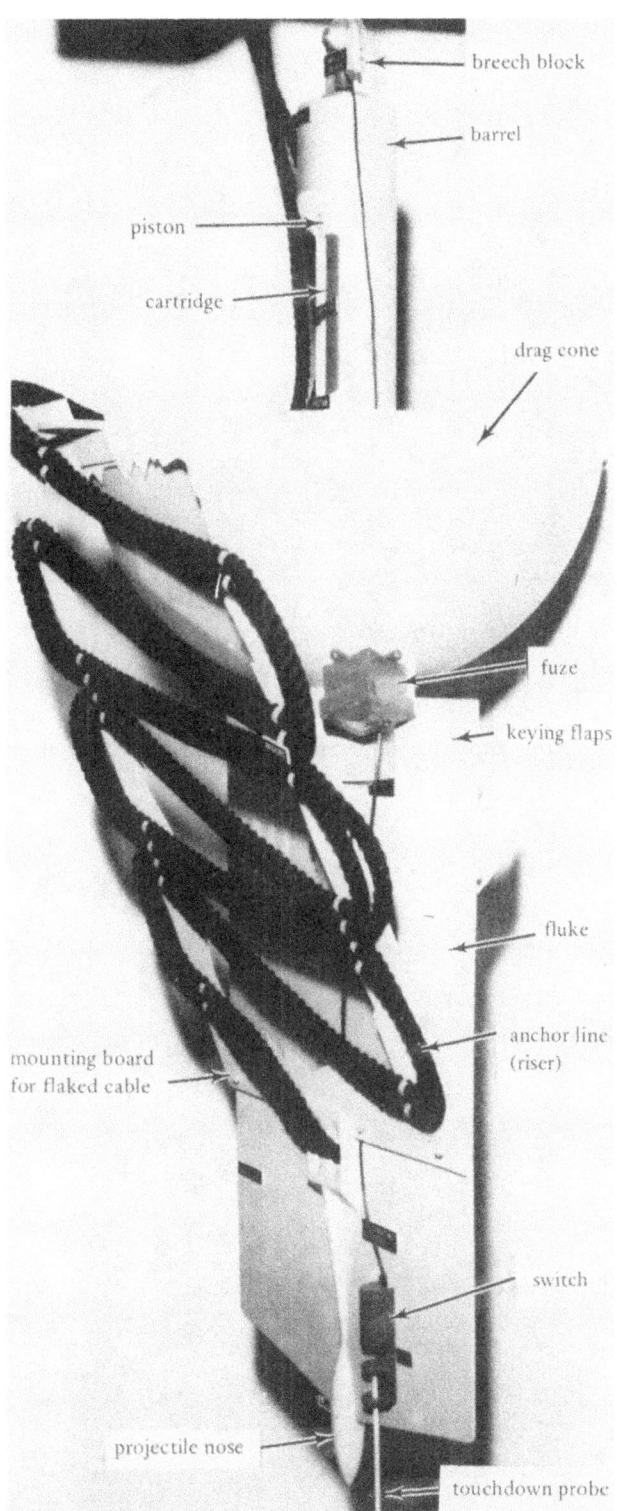

Figure 3.9-1. Army explosive embedment anchor, XM-200; cutaway model.

December 1974

EXPLOSIVE EMBEDMENT ANCHOR, XM-200

Figure 3.9-2. Army explosive embedment anchor, XM-200; front quarter view.

Figure 3.9-3. Army explosive embedment anchor, XM-200; rear quarter view.

PACAN 3DT

3.10. PACAN 3DT (Propellant-Actuated)

3.10.1. Source

Union Industrielle et d'Enterprise
49 bis, Avenue Hache
75008 Paris, France

3.10.2. General Characteristics

An operational anchor whose known testing has been confined to corals and shelly limestone (30 installations). It was designed as a mooring anchor for both sediments and rock (see Figure 3.10-1).

Advertised Nominal Holding Capacity

66,000 lb

Nominal Penetration

—

Water Depth

Design values —
  Maximum: 3,000 ft and 20,000 ft (two designs)
  Minimum: —
Experience —
  Maximum: 300 ft
  Minimum: —

Limitations

—

Advantageous Features

Many expensive components are recoverable and reusable

A special auxiliary connector in the anchor line (optional) is designed to permit recovery of the gun assembly without line entanglement, facilitate emplacement of heavy mooring lines, and permit installation of the mooring some time after installation of the anchor

3.10.3. Details

Anchor Assembly (see Figure 3.10-1)

Height with spike projectile, including probe: 25 ft
Height with plate projectile, including probe: 16 ft (estimated)
Outside diameter: 4.6 ft
Weight, including plate projectile and drum for pendant: 5,300 lb

Anchor-Projectiles

For sand (see Figures 3.10-2 and 3.10-3) —
  Type: Rotating plate of arrowhead shape, with stiffening ribs
  Length of projectile: 4.7 ft
  Length of fluke: 4.7 ft
  Width of fluke: 2.4 ft
  Effective area of fluke: 7.4 sq ft
  Weight: 800 lb

For rock (see Figure 3.10-3) —
  Type: Spike
  Length: 17 ft (estimated)
  Weight: —

Gun Assembly

Barrel diameter (inside): 4 in. (approx)
Length of travel: —
Maximum working pressure: —
Separation velocity: —
Upward reaction distance: —
Propellant: —
Primer: —

First edition—blanks will be eliminated as revisions are made.

December 1974

PACAN 3DT

3.10.4. Operational Aspects

Operational Modes

1. Cable-lowered, automatic-firing, gun assembly recovered (primary mode)
2. Cable-lowered, command-firing, gun assembly recovered

Safety Features

Hydrostatic-pressure arming

Shorting of leads at surface vessel (optional, command-firing mode)

3.10.5. Cost

The material cost per anchor installation when purchased in lots of from 1 to 20 anchors is $4,570. This assumes one reusable gun assembly per 20 firings with the gun assembly costing $17,300. Note: Cost will vary ±$500 per anchor according to type of fluke. Cost figures include 2-1/2% charge for packaging for export. Cost figures pertain to gun designed for 3,000-ft depth; add approximately $5,700 for gun designed for 20,000 ft.

3.10.6. References

1. Letter, P.D.L. (MAREP) to W. J. Tudor (NAVFAC), Jul 22, 1969.

2. Letter, J. Liautaud (UiE) to R. J. Taylor (CEL), Sep 11, 1973.

Figure 3.10-1. PACAN 3DT; equipped with plate-type fluke mounted in cradle aboard ship.

PACAN 3DT

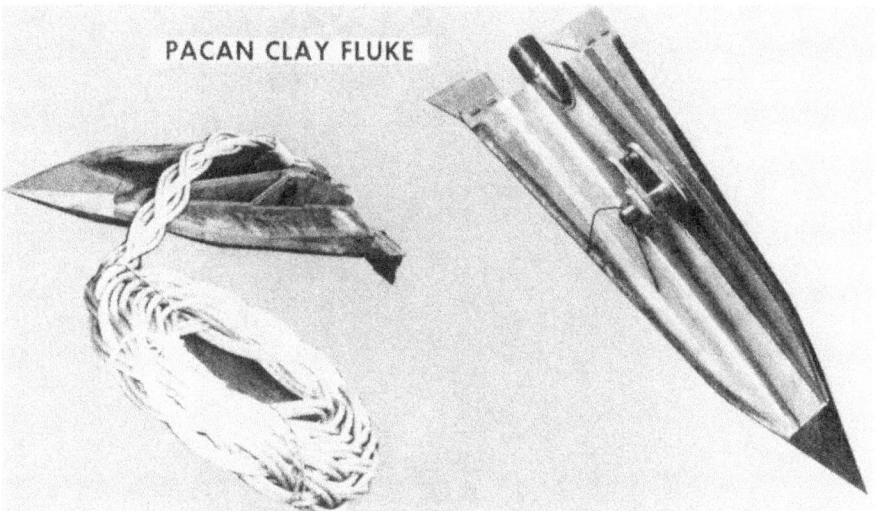

Figure 3.10-2. PACAN 3DT; anchor-projectiles for sediments.

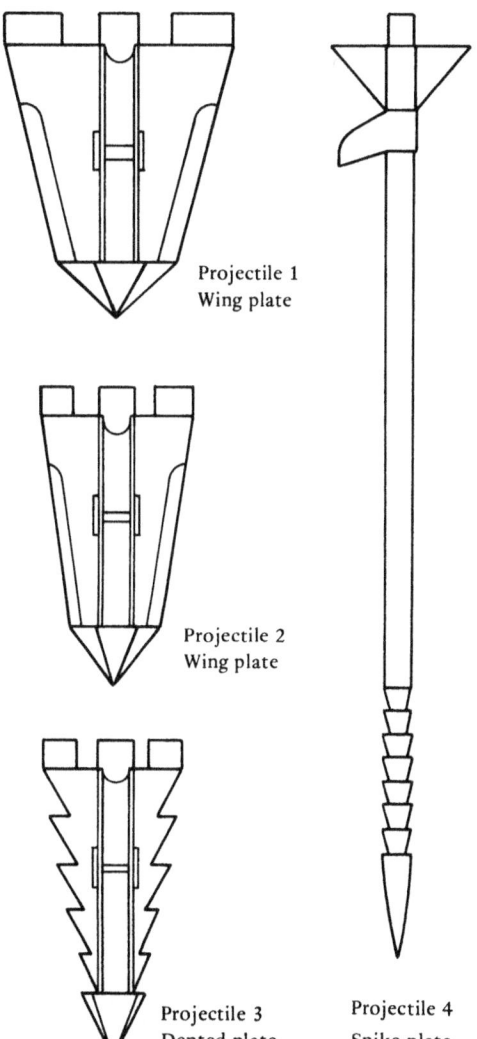

Projectile 1
Wing plate

Projectile 2
Wing plate

Projectile 3
Dented plate

Projectile 4
Spike plate

Figure 3.10-3. PACAN 3DT; anchor-projectiles.

December 1974

3.11. PACAN 10 DT (Propellant-Actuated)

3.11.1. Source

Union Industrielle et d'Enterprise
49 bis, Avenue Hache
75008 Paris, France

3.11.2. General Characteristics

This anchor has been fabricated, but is untested. It was designed as a large-capacity anchor for sediments and rock (see Figure 3.11-1).

Advertised Nominal Holding Capacity

220,000 lb

Nominal Penetration

—

Water Depth

Design values —
Maximum:	3,000 ft
Minimum:	—

Experience —
Maximum:	Not tested
Minimum:	Not tested

Limitations

—

Advantageous Features

Many expensive components are recoverable and reusable

A special auxiliary connector in the anchor line (optional) is designed to permit recovery of the gun assembly without line entanglement, facilitate emplacement of heavy mooring lines, and permit installation of the mooring some time after installation of the anchor

3.11.3. Details

Anchor Assembly (see Figure 3.11-1)

Height with spike projectile, including probe:	44 ft
Height with plate projectile, including probe:	31 ft (estimated
Outside diameter:	7.2 ft
Weight, including plate projectile and drum for pendant:	19,400 lb

Anchor-Projectile

For sand (see Figures 3.10-2 and 3.10-3) —

Type:	Rotating plate of arrowhead shape, with stiffening ribs
Length of projectile:	—
Length of fluke:	9.3 ft
Width of fluke:	3.0 ft
Effective area of fluke:	18 sq ft
Weight:	3,000 lb (estimated)

For rock (see Figure 3.10-3) —

Type:	Spike
Length:	—
Weight:	—

Gun Assembly

Barrel diameter (inside):	8 in. (approx)
Length of travel:	—
Maximum working pressure:	—
Separation velocity:	—
Upward reaction distance:	—
Propellant:	—
Primer:	—

3.11.4. Operational Aspects

Operational Modes

1. Cable-lowered, automatic-firing, gun Assembly recovered (primary mode)
2. Cable-lowered, command-firing, gun assembly recovered

First edition—blanks will be eliminated as revisions are made.

December 1974

PACAN 10 DT

Safety Features

—

3.11.5. Cost

The material cost per anchor installation when purchased in lots of from 1 to 20 anchors is $12,570. This assumes one reusable gun assembly per 20 firings with the gun assembly costing $29,400. Note: Costs are approximate values for plate projectiles. Add approximately $1,300 for spike anchors. Cost figures include 2-1/2% charge for packaging for export.

3.11.6. References

1. Letter, P.D.L. (MAREP) to W. J. Tudor (NAV-FAC) Jul 22, 1969.

2. Letter, J. Liautaud (UiE) to R. J. Taylor (CEL), Sep 11, 1973.

Figure 3.11-1. PACAN 10DT; without anchor projectile.

DIRECT-EMBEDMENT VIBRATORY ANCHOR

3.12. DIRECT-EMBEDMENT VIBRATORY ANCHOR (Vibrated)

3.12.1. Source

Civil Engineering Laboratory
Naval Construction Battalion Center
Port Hueneme, California 93043

3.12.2. General Characteristics

An operational, reliable anchor for direct embedment in all sediments and in water depths to 6,000 ft. It combines low cost (components either inexpensive or recoverable) with lightweight, and it develops holding capacities for loads in any direction.

Advertised Nominal Holding Capacity

Sand:	40,000 lb
Clay:	25,000 lb

Nominal Penetration

Sand:	10 ft
Clay:	20 ft

Water Depth

Design values —
    Maximum: 6,000 ft
    Minimum: 0 ft

Experience —
    Maximum: 6,000 ft
    Minimum: 30 ft

Limitations

Relatively sensitive to wind, seas, and currents during installation, because of the relatively longer time during which the surface vessel must remain on station

Fairly smooth and level seafloor required by the support-guidance frame

Advantageous Features

Expendable components of installation system are relatively inexpensive

Penetration can be monitored and holding capacity predicted without prior investigation

No lines to the surface other than anchor line

3.12.3. Details

Anchor Assembly (see Figure 3.12-1)

Height (based on 15-ft shaft):	19 to 21 ft depending upon size of fluke
Maximum diameter (support-guidance frame):	8 ft
Weight:	1,800 lb

Fluke-Shaft Assembly

Fluke (see Figure 3.12-2) —

Type:	Rotating "Y-fins" (three semi-circular steel plates joined along their straight edges to form a Y-section with 120-degree dihedral angles); upper half of one plate omitted to make room for keying linkage
Fluke diameter:	2, 3, and 4 ft
Plate thickness:	1/2 in.

Keying linkage —

Type:	Two-bar linkage between collar at base of shaft and outer corner of quarter-circle fin; fluke rotates when shaft moves upward

Fluke-shaft locking mechanism (see Figure 3.12-3) —

Type:	Two over-center toggles pinned to shaft, and tension straps from toggles to fluke
Function:	Locks fluke securely to shaft during penetration; released by tripping slug at end of anchor cable inside shaft when upward load is applied to cable

First edition—blanks will be eliminated as revisions are made.

December 1974

DIRECT-EMBEDMENT VIBRATORY ANCHOR

Shaft —
- Type: 3-in. schedule 80 pipe
- Length: 15 ft (normal; readily varied)

Drive Assembly

Vibrator —
- Type: Two counter-rotating masses
- Location: On shafts of motors in housing mounted on upper end of shaft
- Peak force: 12,500 lb at 4,500 rpm

Motor —
- Number: 2
- Type: Electrical (DC)
- Power: 4 hp

Support-Guidance Frame (see Figure 3.12-4)

Base —
- Construction: Welded hexagon of 3-in. pipe
- Outside diameter: 8 ft

Support —
- Type: Tripod of 3-in. pipe; lower ends pinned to base so as to be collapsible, and upper ends fastened to guide-sleeve segments
- Height: 6 ft

Guide sleeve —
- Construction: Three 120-degree portions of a circular cylinder held together by a clamp
- Function: Guides shaft at start of embedment process; proximity of vibrator releases clamp, allows supports to collapse, and permits penetration to continue until entire shaft is embedded

Energy Source
- Type: Lead-acid batteries (twenty 12-volt, 30-amp-hr)
- Life: 60 min, approx (sustained load)
- Location: In three boxes mounted on base of support-guidance frame

3.12.4. Operational Aspects

Operational Modes

1. Shallow water: cable-lowered, surface-powered, used without support guidance frame, drive assembly not recovered
2. Deep water: Cable-lowered, automatic-starting, uncontrolled power supply, drive assembly and supports, guidance frame not recovered

Safety Features

Accommodates standard field safety practice

3.12.5. Cost

Shallow water (<300 ft): $4,000 (approx) per placement. Deep water: $10,000 (approx) per placement.

3.12.6. References

1. Naval Civil Engineering Laboratory. Contract Report No. CR 69-009: Vibratory embedment anchor system. Long Beach, CA, Ocean Science and Engineering, Inc., Feb 1969. (Contract no. N62399-68-C-0008) (AD848920L)

2. Naval Civil Engineering Laboratory. Technical Note N-1133: Specialized anchors for the deep sea; progress summary, by J. E. Smith, R. M. Beard, and R. J. Taylor. Port Hueneme, CA, Nov 1970. (AD716408)

DIRECT-EMBEDMENT VIBRATORY ANCHOR

3. Naval Civil Engineering Laboratory. Technical Report R-791: Direct embedment vibratory anchor, by R. M. Beard. Port Hueneme, CA, Jun 1973. (AD766103)

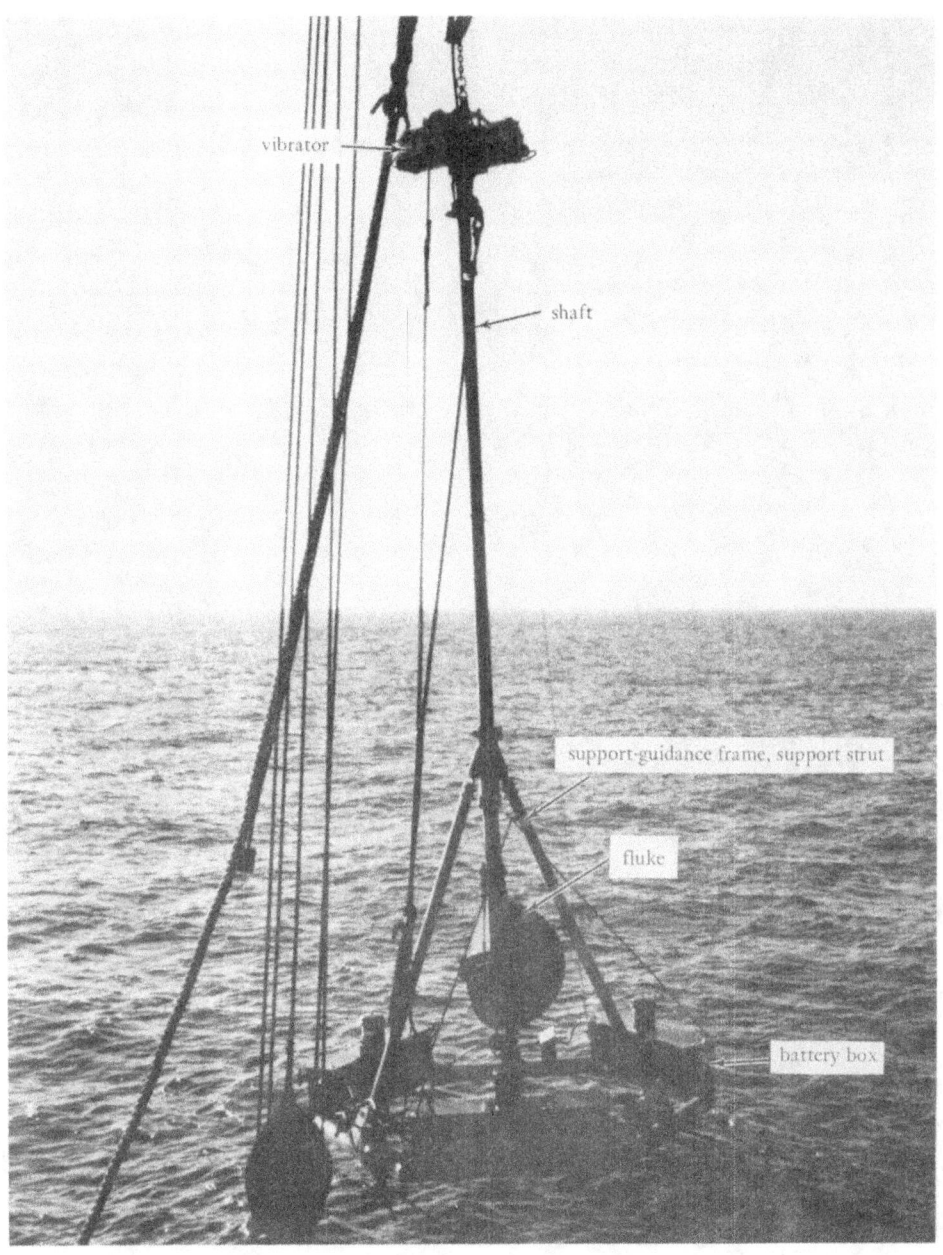

Figure 3.12-1. Navy vibratory anchor.

DIRECT-EMBEDMENT VIBRATORY ANCHOR

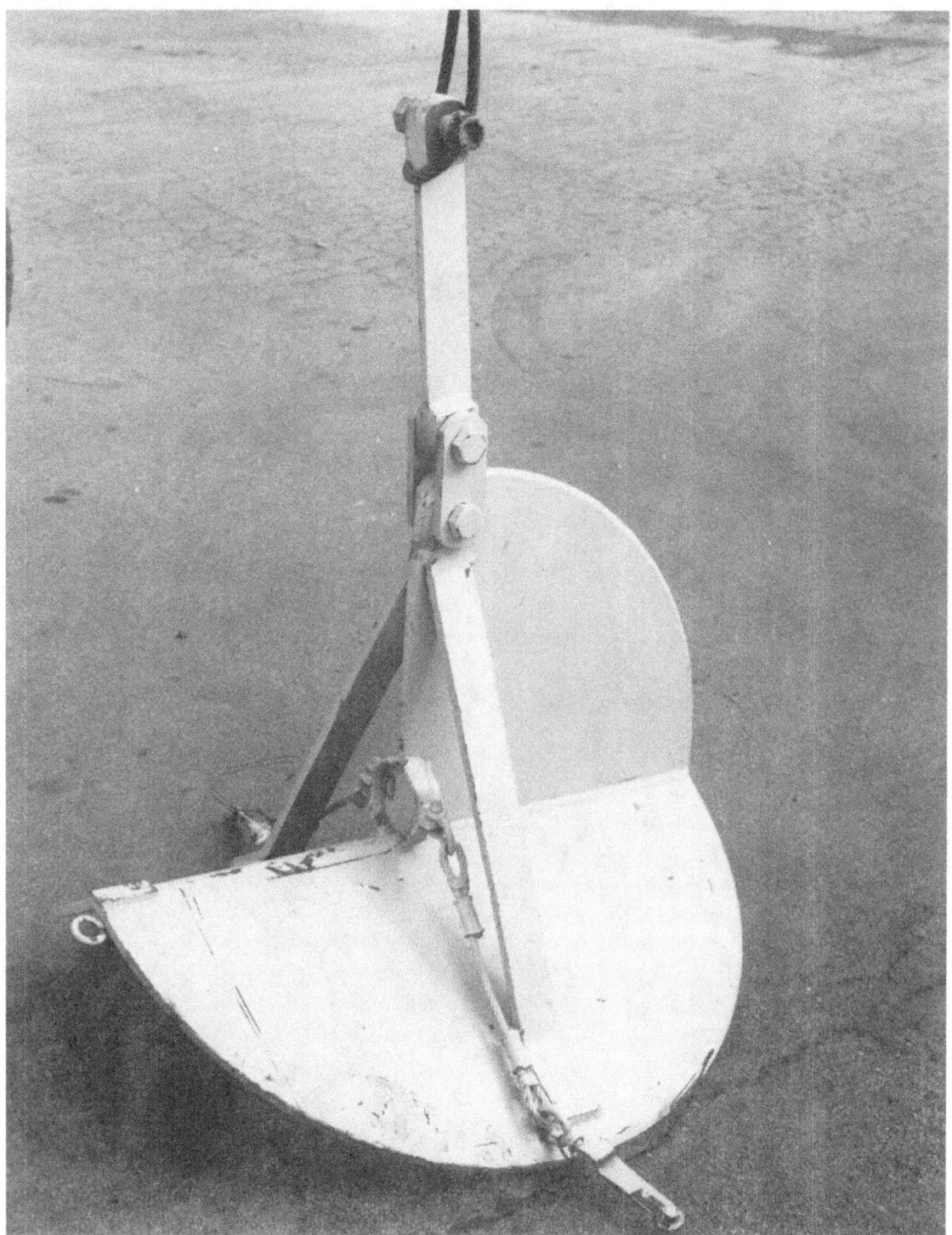

Figure 3.12-2. Navy vibratory anchor; quick-keying fluke shown in position assumed after keying.

December 1974

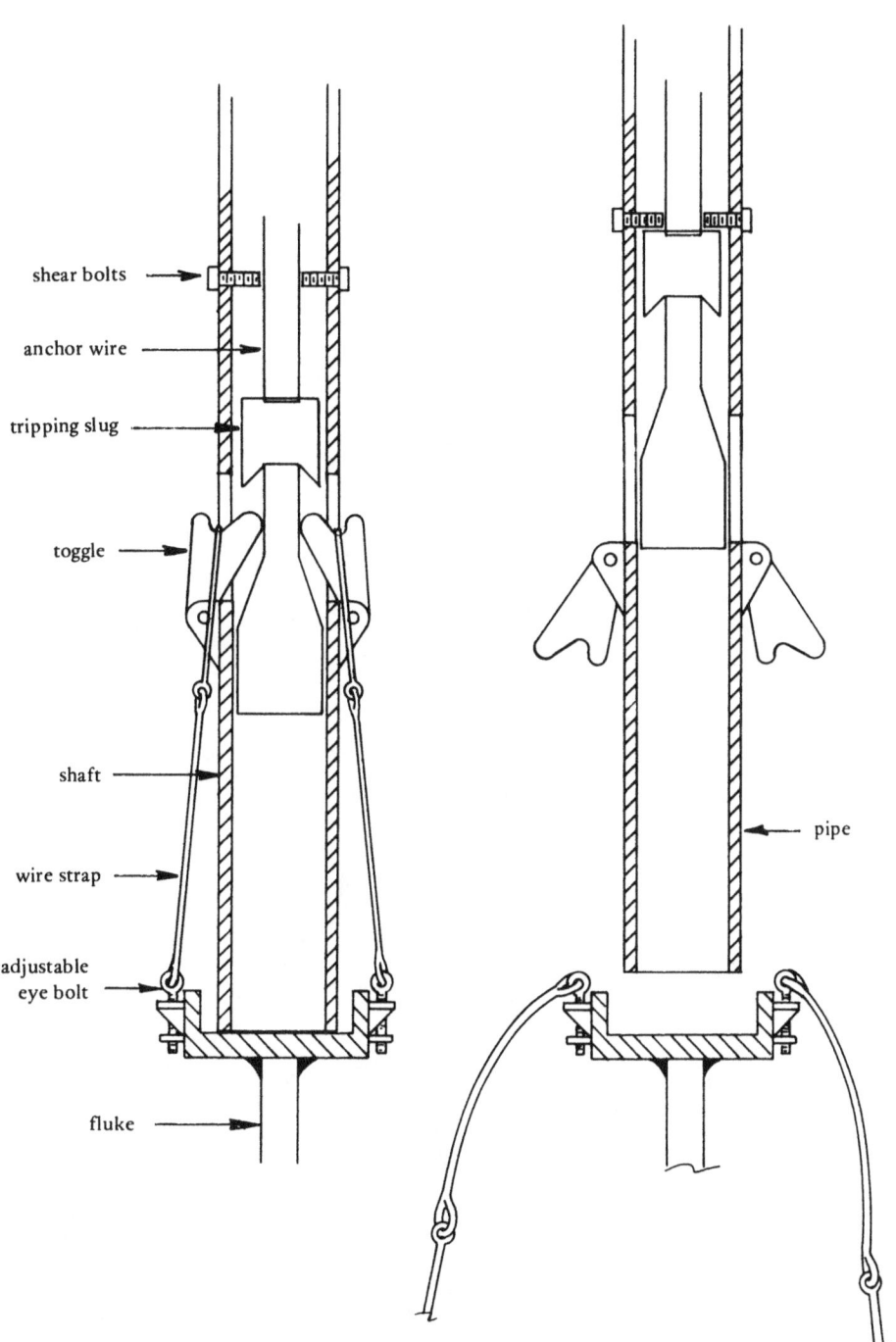

Figure 3.12-3. Navy vibratory anchor; fluke locking mechanism.

December 1974

DIRECT-EMBEDMENT VIBRATORY ANCHOR

Figure 3.12-4. Navy vibratory anchor; embedded in sand on beach to demonstrate collapsible support-guidance frame.

December 1974

VIBRATORY EMBEDMENT ANCHOR, MODEL 2000

3.13. VIBRATORY EMBEDMENT ANCHOR, MODEL 2000 (Vibrated)

3.13.1. Source

Ocean Science and Engineering, Inc.
5541 Nicholson Lane
Rockville, Maryland 20852

3.13.2. General Characteristics

An operational, low-cost, lightweight, fairly high capacity anchor for sediments. It is used in shallow-to-moderate depth (500 ft) for taut-line tethers, short-scope ship moorings, and other applications requiring precise placement of the anchor (see Figure 3.13-1).

Advertised Nominal Holding Capacity

80,000 lb

Nominal Penetration

40 ft

Water Depth

Design values —	
Maximum:	500 ft
Minimum:	5 ft
Experience —	
Maximum:	—
Minimum:	—

Limitations

Relatively sensitive to wind, seas, and currents during installation, because of the relatively longer time during which the surface vessel must remain on station

Potential for entanglement of multiple lines

Advantageous Features

Most of the installation equipment is recoverable and reusable

Penetration can be monitored and holding capacity predicted without prior investigation

3.13.3. Details

Anchor Assembly (see Figure 3.13-1)

Height (based on 40-ft shaft):	43 ft (approx)
Maximum transverse dimension (tether bar):	7 ft
Weight:	1,000 lb

Fluke Assembly (see Figures 3.13-1 and 3.13-2)

Fluke —

Type:	Rotating "Y-fins" (three semi-circular steel plates joined along their straight edges to form a Y-section with 120-degree dihedral angles); upper half of one plate omitted to make room for keying linkage
Fluke diameter:	3 ft
Plate thickness:	3/8 in.

Keying linkage —

Type:	Two-bar linkage between collar at base of shaft and outer corner of quarter-circle fin; fluke rotates when shaft move upward

Shank Assembly

Shaft —

Construction:	4-in. schedule 40 pipe
Length:	40 ft (normal; readily varied)

Tension member —

Construction:	3/4-in. wire rope inside the shaft, extending from the fluke to a tensioning device at the upper end of the shaft

First edition—blanks will be eliminated as revisions are made.

December 1974

VIBRATORY EMBEDMENT ANCHOR, MODEL 2000

Function:	Secures the fluke, shaft, and drive assembly together during penetration	Capacity: Pressure:	0 to 27 gpm 0 to 3,000 psi

Prime mover —

Type:	Diesel engine
Location:	On surface vessel
Size:	6-cylinder, 100-hp

Tensioning device —

Construction:	Hand-operated, 3,000-psi hydraulic cylinder by which collar on upper end of tension member is pulled upward
Location:	Attached to vibrator housing

Tether bar —

Construction:	Steel bar, 7 ft long, pin-connected to collar near upper end of shaft; collar swivels around shaft
Function:	Point of attachment of anchor line; permits swinging of moored vessel

Drive Assembly

Vibrator —

Type:	Two counter-rotating masses
Location:	On shafts of motors in housing mounted on upper end of shaft
Peak force:	24,000 to 30,000 lb at 3,600 rpm

Motor —

Number:	2
Type:	Hydraulic
Capacity:	17 gpm at 3,600 rpm

Power Source

Pump —

Type:	Hydraulic, variable positive-displacement
Location:	On surface vessel

3.13.4. Operational Aspects

Operational Modes

Cable-lowered, remote-manual starting, remote-manual control of power supply, drive assembly recovered

Safety Features

Accommodates standard field safety practice

3.13.5. Cost

The material cost per anchor installation when purchased in lots of from 1 to 50 anchors is $3,184. This assumes one reusable drive assembly per 100 installations with the drive assembly costing $23,400.

3.13.6. References

1. S. H. Shaw. "New anchoring concept moors floating drydock," Ocean Industry, vol 7, no. 1, Jan 1972, pp 31-33.

2. Letter, R. L. Fagan (OS&E) to R. J. Taylor (CEL), 11 Dec 1973.

3. Ocean Science and Engineering, Inc. Pamphlet: New anchoring system: Vibratory embedment anchor, Model 2000, Rockville, MD, undated.

VIBRATORY EMBEDMENT ANCHOR, MODEL 2000

Figure 3.13-1. Ocean Science and Engineering vibratory embedment anchor, Model 2000.

December 1974

CHANCE SPECIAL OFFSHORE MULTI-HELIX SCREW ANCHOR

3.14. CHANCE SPECIAL OFFSHORE MULTI-HELIX SCREW ANCHOR (Screw-In)

3.14.1. Source

Anchoring Inc.
P.O. Box 55263
Houston, Texas 77055

3.14.2. General Characteristics

An operational, reliable anchor for use primarily in sediments and in moderately shallow water. It can be installed rapidly and precisely, and it is used extensively for pipeline tiedown (See Figure 3.14-1).

Advertised Nominal Capacity

10,000 lb

Nominal Penetration

10 ft

Water Depth

Design values —
  Maximum: —
  Minimum: —

Experience —
  Maximum: 325 ft
  Minimum: 0 ft

Limitations

For precise placement, relatively quiet conditions (wind, wave, current) are required during positioning of the drive system and initial phase of embedment

Advantageous Features

Several simple options are available for increasing holding capacity: diameter, number, and spacing of helixes; torsional strength of shaft; and depth of penetration

3.14.3. Details

Anchor (see Figure 3.14-2)

Type:	Two to four circular, single-turn, helical surfaces spaced along a circular shaft (square shafts available)
Shaft outside diameter:	1-1/4 in. (larger sizes available)
Shaft length —	
Anchor section (carries helixes):	5, 7, or 10 ft for two, three, or four helixes, respectively
Extension section (no helixes):	10 ft maximum
Helix diameter:	4 or 6 in. (larger sizes available)
Weight:	100 lb (average)

Installation Assembly (Pipeline Anchors) (see Figures 3.14-1 and 3.14-3)

Size:	Approx 3 x 5 x 8 ft without anchors
Weight:	6,000 lb
Drive heads —	
Type:	Two counter-rotating, gear-driven
Speed:	45 rpm
Maximum torque:	Greater than 5,000 ft-lb
Motors —	
Number:	2
Type:	Hydraulic
Flow rate:	25 to 30 gpm
Buoyancy tank:	—

Power System

Location:	On support vessel
Size:	Approx 5 x 5 x 5 ft
Pump —	
Flow rate:	25 to 30 gpm
Maximum pressure:	2,000 psi

First edition—blanks will be eliminated as revisions are made.

December 1974

CHANCE SPECIAL OFFSHORE MULTI-HELIX SCREW ANCHOR

Prime mover: Diesel engine
Air compressor: —

3.14.4. Operational Aspects

Operational Modes (Pipeline Anchors)

1. Cable-lowered, position controlled by divers, installation assembly recovered
2. Cable-lowered, position controlled by television, installation assembly recovered

Safety Features

Accommodates standard field safety practice

3.14.5. Cost

Approximately $375 per pair, installed. Anchoring Inc. installs all the anchors.

3.14.6. References

1. A. B. Chance Co. Bulletin 424-C: No-wrench screw anchors, Centralia, MO, 1969. (Part C of Encyclopedia of Anchoring)

2. G. E. Cannon. "Pipe anchors pin line solidly to sea floor," Offshore, vol 29, no. 12, Nov 1969, pp 84, 86.

3. Letter, G. E. Cannon (Anchoring, Inc.) to R. J. Taylor (CEL), Sep 4, 1973.

Figure 3.14-1. Chance Special Offshore Multi-Helix system for pipeline anchoring; pipeline bracket visible.

CHANCE SPECIAL OFFSHORE MULTI-HELIX SCREW ANCHOR

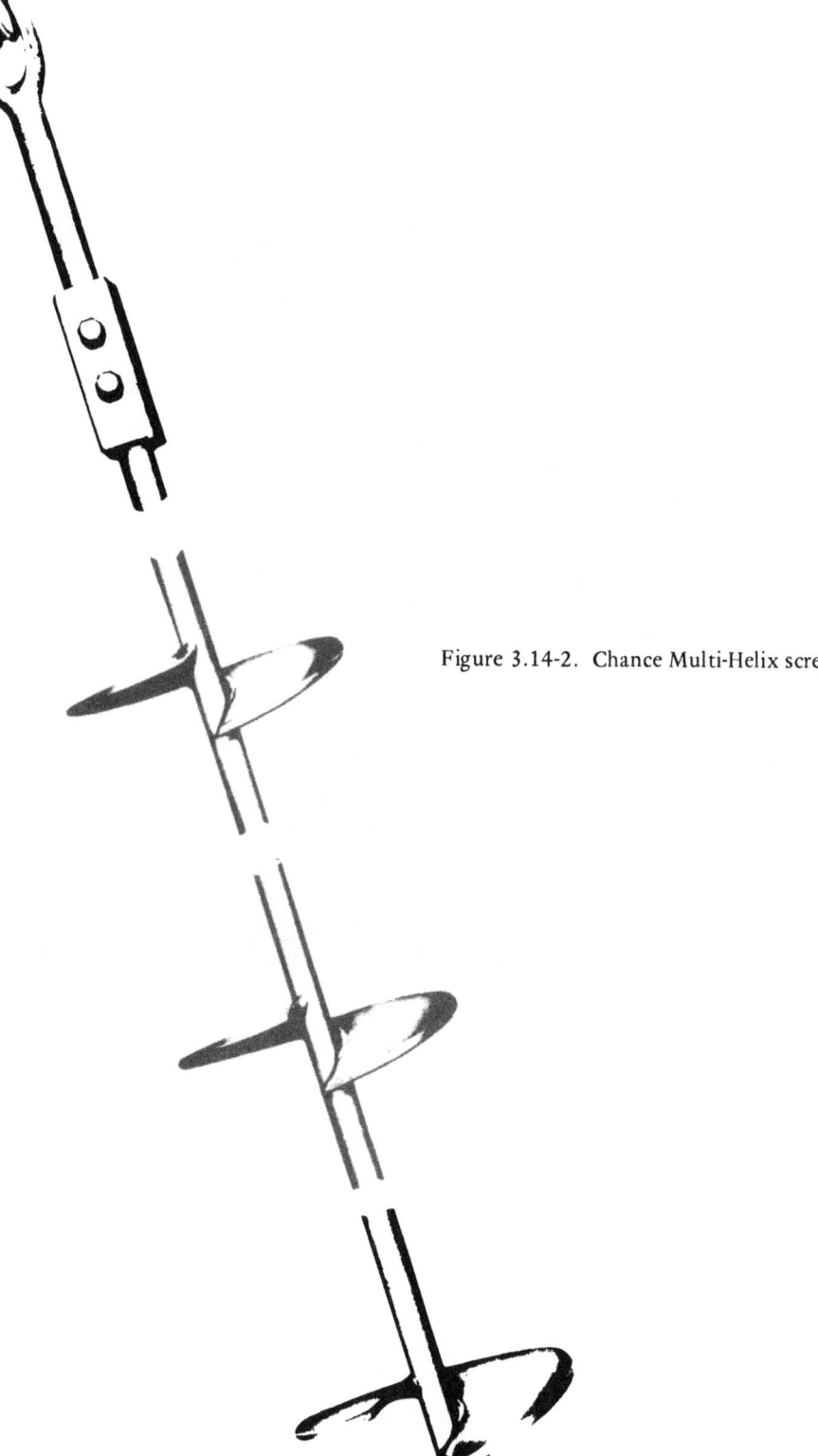

Figure 3.14-2. Chance Multi-Helix screw anchor.

CHANCE SPECIAL OFFSHORE MULTI-HELIX SCREW ANCHOR

Figure 3.14-3. Chance Special Offshore Multi-Helix system for pipeline anchoring.

STAKE PILE

3.15. STAKE PILE (Driven)

3.15.1. Source

Naval Facilities Engineering Command
200 Stovall Street
Alexandria, Virginia 22332

3.15.2. General Characteristics

An operational anchor that has been tested and used in East Coast locations to secure mothball fleets. It comprises a family of moderate-to-large-capacity anchors for Fleet moorings for ships and floating drydocks that will not drag under load and does not require dragging for setting.

Advertised Nominal Holding Capacity

For 8-in. pile size —

Sand:	100,000 lb
Soft clay:	20,000 lb

For 12-in. pile size —

Sand:	200,000 lb
Soft clay:	30,000 lb

For 16-in. pile size —

Sand:	300,000 lb
Soft clay:	40,000 lb

Nominal Penetration

35 ft (top of pile 5 ft below firm bottom)

Water Depth

Design values —

Maximum:	Determined largely by available pile-driving equipment
Minimum:	0 ft

Experience —

Maximum:	—
Minimum:	0 ft

Limitations

Horizontal component of load on the pile is desirable

Maximum depth of water determined by the pile-driving equipment available

Not efficient in soft clay

Advantageous Features

Simple structure

3.15.3. Details

Anchor (see Figures 3.15-1 and 3.15-2)

Description:	Steel tubes, 30 ft long, with four fins extending along the upper 14 ft

Outside diameter —

For 8-in. pipe:	8.75 in.
For 12-in. pipe:	12.75 in.
For 16-in. pipe:	16.00 in.

Pipe wall thickness —

For 8-in. pipe:	0.25 in.
For 12-in. pipe:	0.25 in.
For 16-in. pipe:	0.375 in.

Width of fins —

For 8-in. pipe:	7 in.
For 12-in. pipe:	10 in.
For 16-in. pipe:	10 in.

Weight —

For 8-in. pipe:	1,400 lb
For 12-in. pipe:	2,600 lb
For 16-in. pipe:	3,600 lb

3.15.4. Operational Aspects

Operational Modes

1. Surface driving
2. Underwater driving

Safety Features

Accommodates standard field safety practice

First edition—blanks will be eliminated as revisions are made.

December 1974

STAKE PILE

3.15.5. Cost

For 8-in. size: $2,500 ea (approx)
For 12-in. size: $3,100 ea (approx)
For 16-in. size: $3,600 ea (approx)

Costs are for hardware only; offshore pile driving currently costs $10,000 to $15,000 per day.

3.15.6. References

1. Naval Civil Engineering Laboratory. Technical Note N-205: Stake pile development for moorings in sand bottoms, by J. E. Smith. Port Hueneme, CA, Nov 1954. (AD81261)

2. Naval Civil Engineering Laboratory. Letter Report L-022: Stake pile tests in mud bottom, by J. E. Smith. Port Hueneme, CA, Sep 1957.

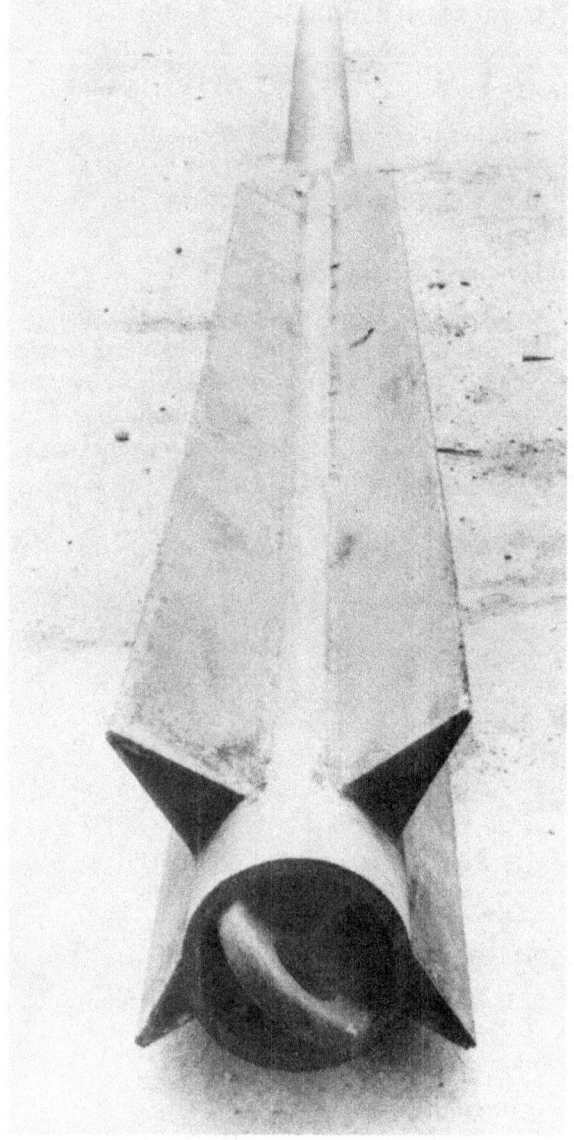

Figure 3.15-1. Navy stake pile; 8-inch.

First edition—blanks will be eliminated as revisions are made.   December 1974

STAKE PILE

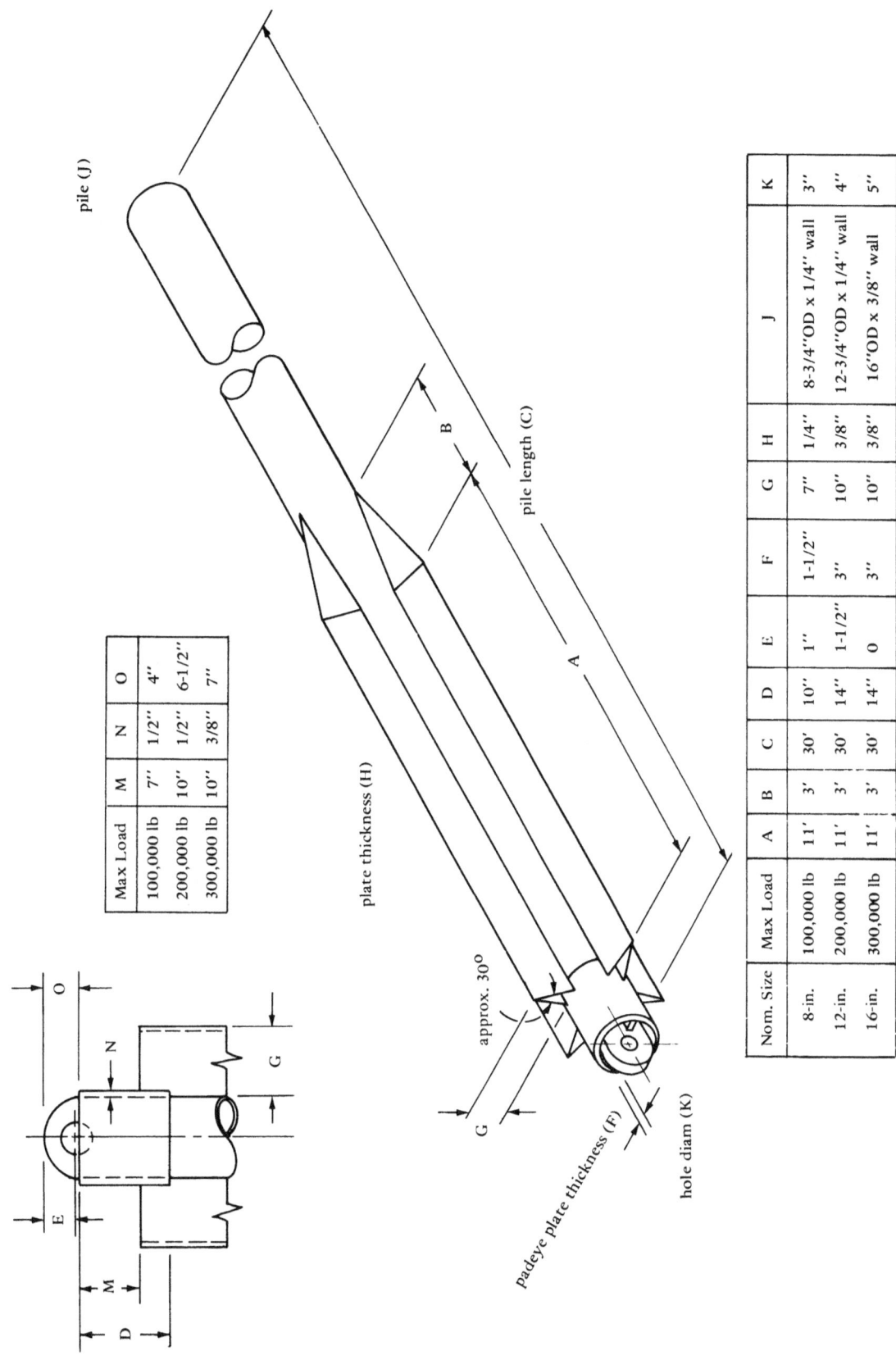

Max Load	M	N	O
100,000 lb	7"	1/2"	4"
200,000 lb	10"	1/2"	6-1/2"
300,000 lb	10"	3/8"	7"

Nom. Size	Max Load	A	B	C	D	E	F	G	H	J	K
8-in.	100,000 lb	11'	3'	30'	10"	1"	1-1/2"	7"	1/4"	8-3/4"OD x 1/4" wall	3"
12-in.	200,000 lb	11'	3'	30'	14"	1-1/2"	3"	10"	3/8"	12-3/4"OD x 1/4" wall	4"
16-in.	300,000 lb	11'	3'	30'	14"	0	3"	10"	3/8"	16"OD x 3/8" wall	5"

Figure 3.15-2. Design specifications for Navy stake pile.

First edition—blanks will be eliminated as revisions are made.

December 1974

UMBRELLA PILE-ANCHOR, MARK III

3.16. UMBRELLA PILE-ANCHOR, MARK III
(Driven)

3.16.1. Source

Naval Facilities Engineering Command
200 Stovall Street
Alexandria, Virginia 22332

3.16.2. General Characteristics

This item has been tested, but not used. It is an anchor for moorings for large vessels which (1) will not drag under load, (2) does not require dragging for pre-setting, and (3) has high capacity in bearing and in resistance to uplift.

Advertised Nominal Holding Capacity

Sand:	300,000 lb

Nominal Penetration

20 ft

Water Depth

Design values –	
Maximum:	Determined largely by available pile-driving equipment
Minimum:	0 ft
Experience –	
Maximum:	Not tested offshore
Minimum:	0 ft

Limitations

Maximum depth of water determined by the pile-driving equipment available

Use restricted to homogeneous, uncemented soils free of boulders and other obstructions

Design not well-adapted to development of a family of anchors of varying size

Advantageous Features

Large capacity in both bearing and resistance to uplift in sand and cohesive soil

3.16.3. Details

Fluke Assembly (see Figures 3.16-1 and 3.16-2)

Type:	Expanding finger type (four flukes)
Length of fluke:	52 in.
Width of fluke:	10 in.
Effective area of flukes:	10.5 sq ft
Angle of rotation from fully closed to fully opened positions:	60 deg
Outside diameter (foot circle of open flukes):	8 ft
Height of assembly:	10 ft
Weight of assembly:	1,400 lb

Chain

Size:	2-3/4-in.
Length:	See length of follower

Follower

Construction:	Steel tubing
Outside diameter:	12.75 in.
Length:	Varies with water depth and embedment depth

Casing

Construction:	Steel tubing
Outside diameter:	18.0 in.
Length:	Varies with water depth and embedment depth

3.16.4. Operational Aspects

Operational Modes

1. Surface driving
2. Underwater driving

First edition–blanks will be eliminated as revisions are made.

December 1974

UMBRELLA PILE-ANCHOR, MARK III

Safety Features

Accommodates standard field safety practice

3.16.5. Cost

$4,500 (approx) per anchor unit. Cost is for hardware only; offshore pile driving currently costs $10,000 to $15,000 per day.

3.16.6. References

1. Naval Civil Engineering Laboratory. Technical Report R-247: Umbrella pile-anchors, by J. E. Smith. Port Hueneme, CA, May 1963. (AD408-404)

Figure 3.16-1. Navy umbrella pile-anchor, Mark III; after test in sand.

UMBRELLA PILE-ANCHOR, MARK III

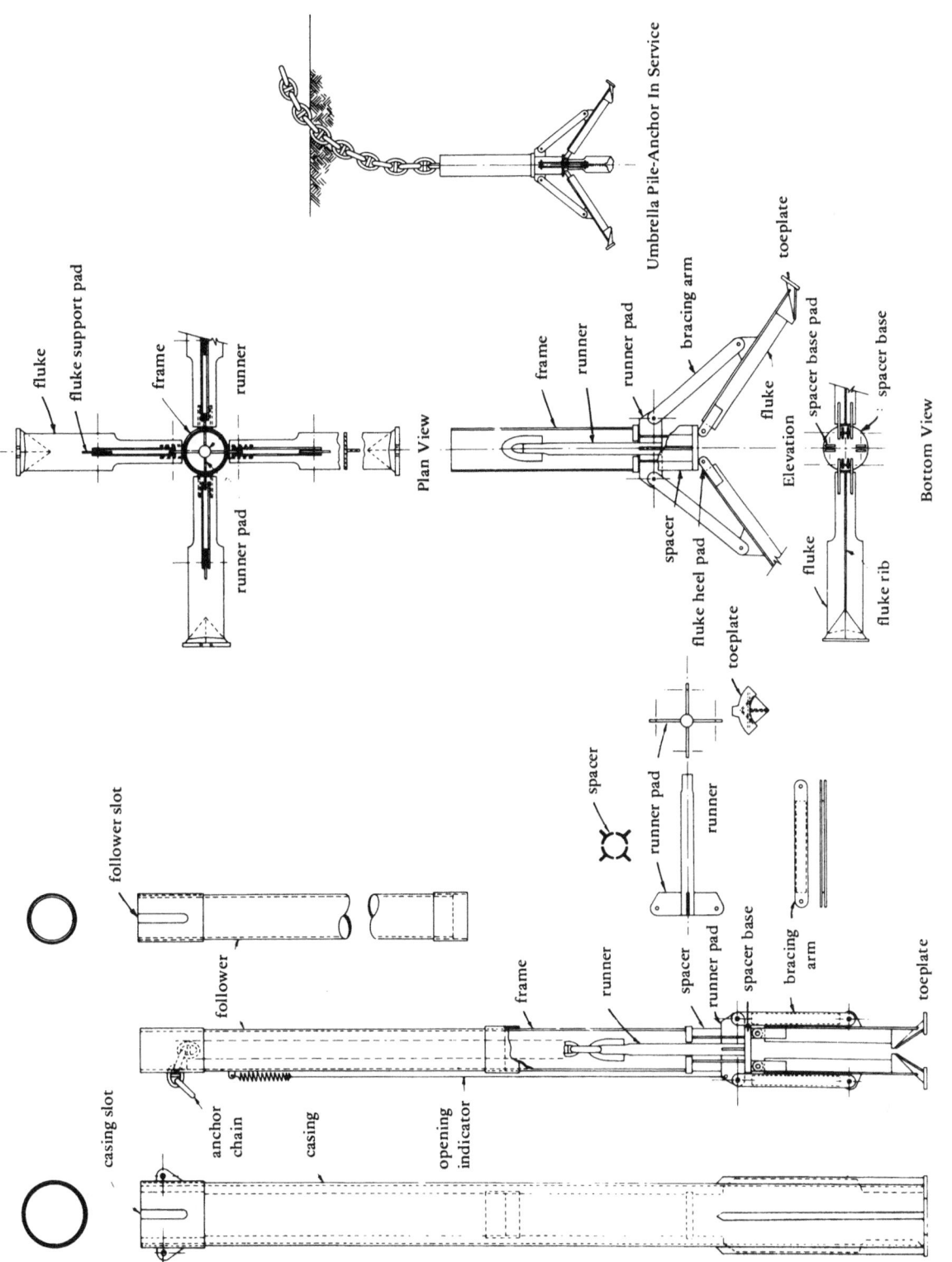

Figure 3.16-2. Navy umbrella pile-anchor, Mark III.

December 1974

UMBRELLA PILE-ANCHOR, MARK IV

3.17. UMBRELLA PILE-ANCHOR, MARK IV
(Driven)

3.17.1. Source

Naval Facilities Engineering Command
200 Stovall Street
Alexandria, Virginia 22332

3.17.2. General Characteristics

This item has been tested but not used. It is an anchor for moorings for large vessels which (1) will not drag under load, (2) does not require dragging for pre-setting, and (3) has high capacity in bearing and in resistance to uplift.

Advertised Nominal Holding Capacity

Sand:	300,000 lb
Mud:	100,000 lb

Nominal Penetration

20 ft

Water Depth

Design values —		
Maximum:		Determined largely by available pile-driving equipment
Minimum:		0 ft
Experience —		
Maximum:		35 ft
Minimum:		0 ft

Limitations

Maximum depth of water determined by the pile-driving equipment available

Use restricted to homogeneous, uncemented soils free of boulders and other obstructions

Advantageous Features

Large capacity in both bearing and resistance to uplift

Functional in sand and cohesive sediments

3.17.3. Details

Fluke Assembly (see Figures 3.17-1 and 3.17-2)

Type:	Expanding finger type (four flukes)
Length of fluke:	49 in.
Width of fluke:	14 in.
Effective area of fluke:	16.5 sq ft
Angle of rotation from fully closed to fully opened position:	60 deg
Outside diameter (foot circle of open flukes):	8 ft
Height of assembly:	8 ft
Weight of assembly:	2,200 lb

Inner Follower

Construction:	Steel tubing
Outside diameter:	12.75 in.
Length:	Varies with water depth and embedment depth

Outer Follower

Construction:	Steel tubing
Outside diameter:	16.0 in.
Length:	Varies with water depth and embedment depth

Chain

Size:	2-3/4 in.
Length:	See length of inner follower

3.17.4. Operational Aspects

Operational Modes

1. Surface driving
2. Underwater driving

First edition—blanks will be eliminated as revisions are made.

December 1974

UMBRELLA PILE-ANCHOR, MARK IV

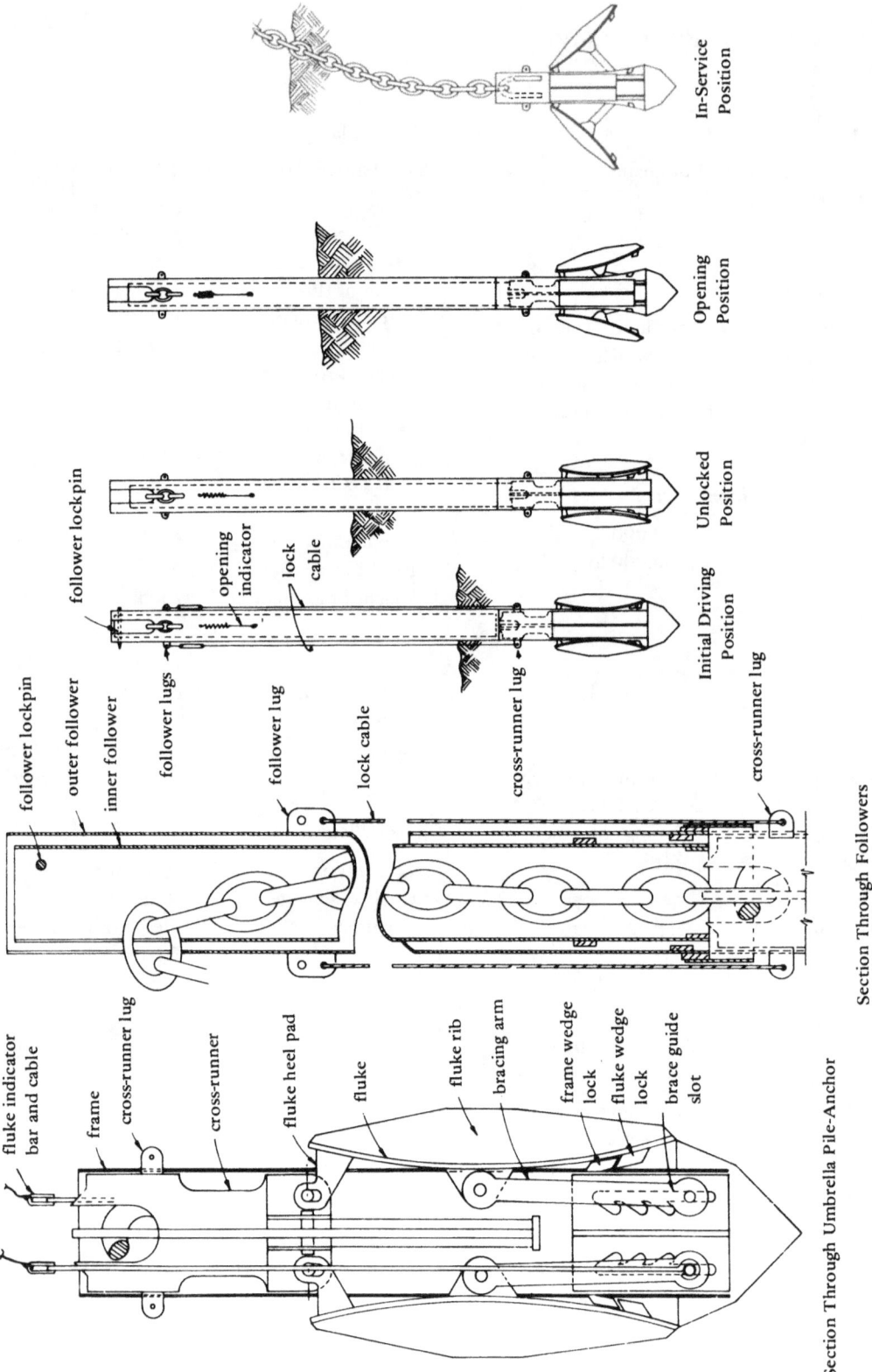

Figure 3.17-1. Navy umbrella pile-anchor, Mark IV.

December 1974

Safety Features

Accommodates standard field safety practice

3.17.5. Cost

$7,500 (approx) per anchor unit. Cost is for hardware only; offshore pile driving currently costs $10,000 to $15,000 per day.

3.17.6. References

1. Naval Civil Engineering Laboratory. Technical Report R-247: Umbrella pile-anchors, by J. E. Smith. Port Hueneme, CA, May 1963. (AD408404)

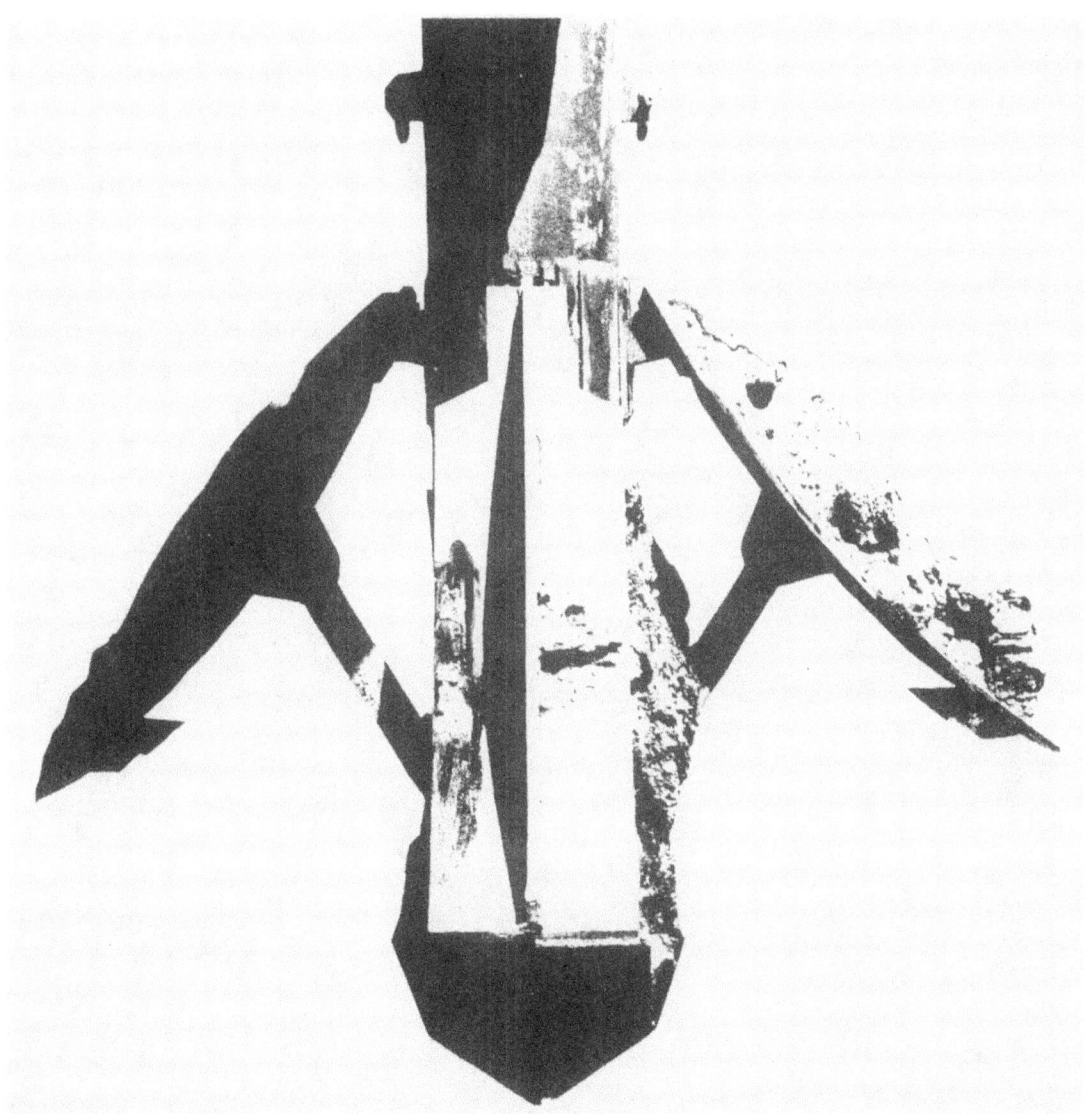

Figure 3.17-2. Navy umbrella pile-anchor, Mark IV; after test in sand.

UMBRELLA PILE-ANCHOR, MARK IV

3.18. ROTATING PLATE ANCHOR (Driven)

3.18.1. Source

Techniques Louis Menard
Centre d'Etudes Geotechniques
Boite Postale No. 2
91 Longjumeau, France

3.18.2. General Characteristics

An operational, high-capacity embedment anchor in sediments for single-point moorings, anchorings in offshore oil operations, and other applications.

Advertised Nominal Holding Capacity

200,000 lb

Nominal Penetration

10 ft to 30 ft

Water Depth

Design values —
  Maximum: —
  Minimum: —

Experience —
  Maximum: —
  Minimum: —

Limitations

Maximum depth of water determined by the pile-driving equipment available

Advantageous Features

—

3.18.3. Details

Fluke (see Figure 3.18-1)

—

Driving Mandrel

—

Chain

—

3.18.4. Operational Aspects

Operational Modes

1. Surface driving
2. Underwater driving

Safety Features

—

3.18.5. Cost

—

3.18.6. References

1. Techniques Louis Menard. Publication P/95: Mooring Anchors. Longjumeau, France, 1970.

UMBRELLA PILE-ANCHOR, MARK IV

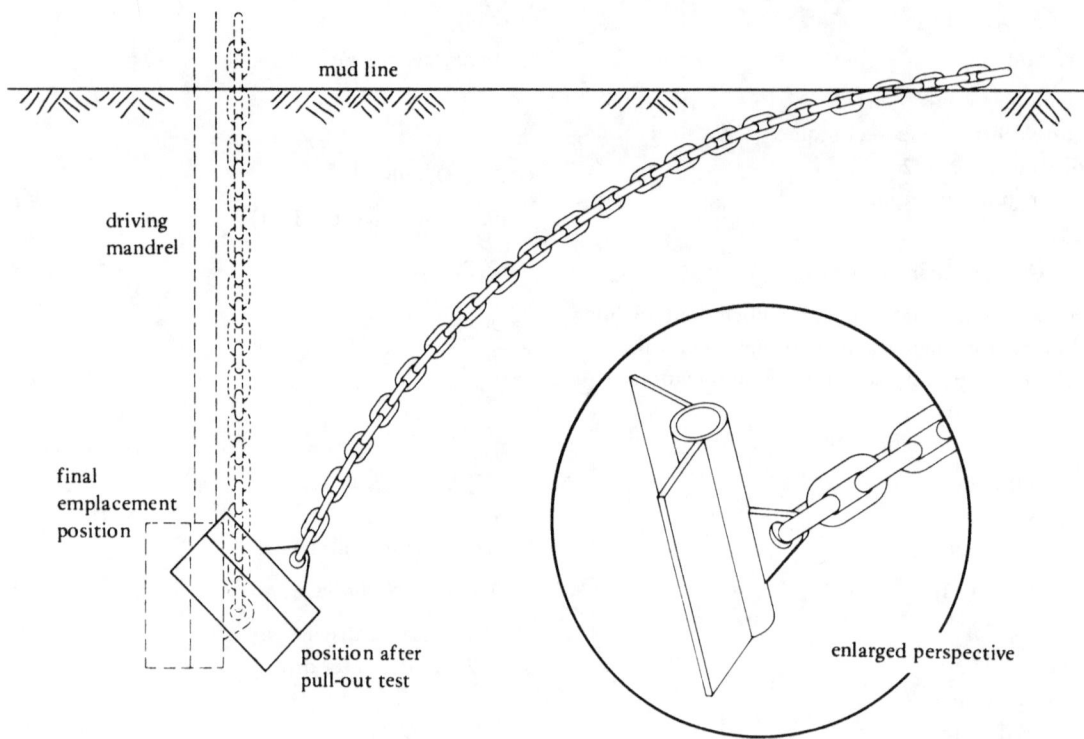

Figure 3.18-1. Menard rotating plate anchor.

EXPANDED ROCK ANCHOR

3.19. EXPANDED ROCK ANCHOR (Drilled)

3.19.1. Source

Techniques Louis Menard
Centre d'Etudes Geotechniques
Boite Postale No. 2
91 Longjumeau, France

3.19.2. General Characteristics

An operational, high-capacity anchor in rock for single-point moorings, anchorings for offshore oil operations, and other applications (see Figure 3.19-1).

Advertised Nominal Holding Capacity

800,000 lb

Nominal Penetration

Rock: 30 ft

Water Depth

0 to 700 ft

Limitations

Relatively long installation time

Advantageous Features

—

3.19.3. Details

—

3.19.4. Operational Aspects

—

3.19.5. Cost

—

3.19.6. References

1. Techniques Louis Menard. Publication P/95: Mooring Anchors. Longjumeau, France, 1970.

First edition—blanks will be eliminated as revisions are made.   December 1974

EXPANDED ROCK ANCHOR

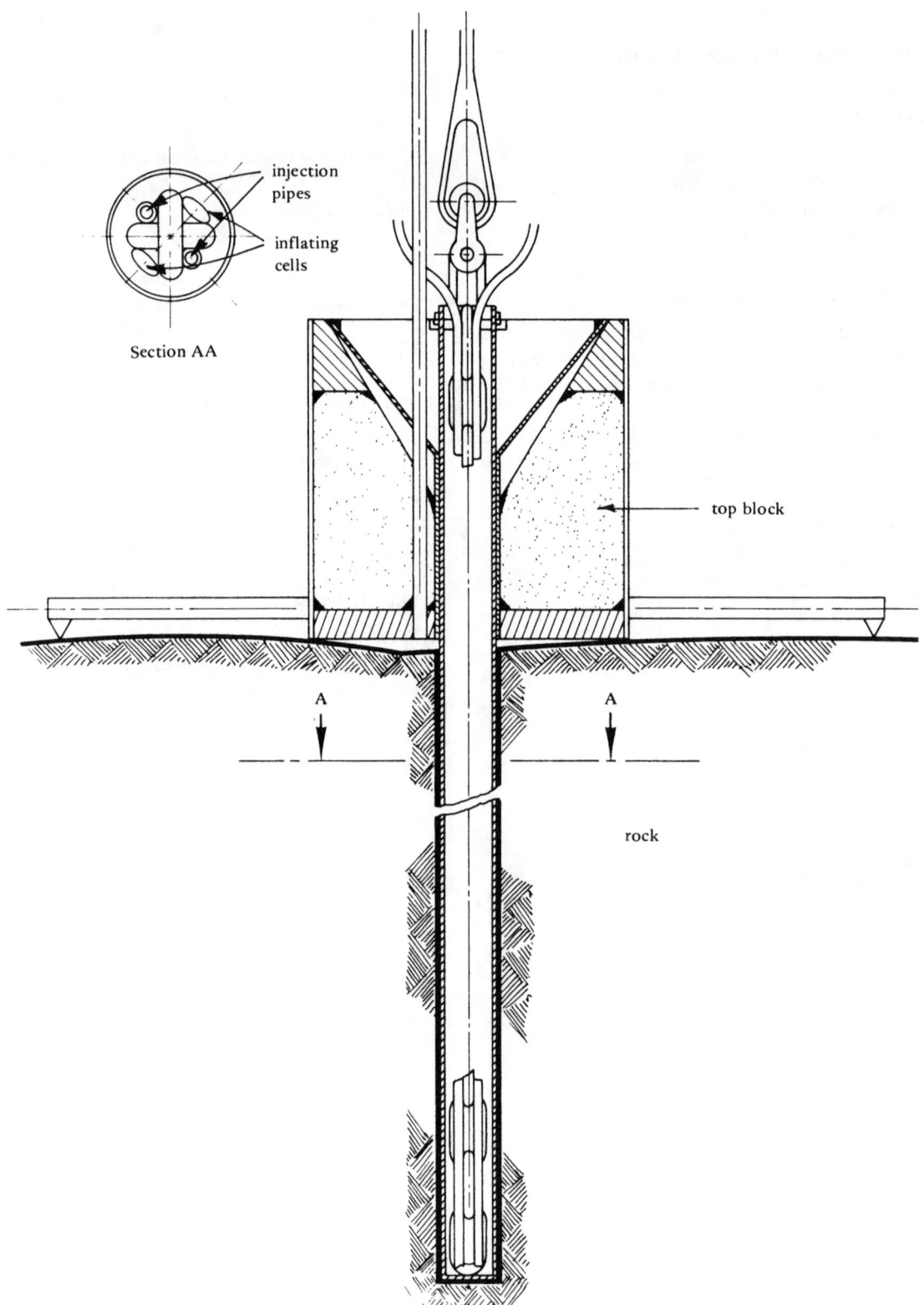

Figure 3.19-1. Menard expanded rock anchor; placement of chain into drilled hole.

December 1974

FREE-FALL ANCHOR SYSTEM

3.20. FREE-FALL ANCHOR SYSTEM (Deadweight)

3.20.1. Source

Delco Electronics
General Motors Corporation
6767 Hollister Avenue
Goleta, California 93017

3.20.2. General Characteristics

An operational item used in numerous moorings (small ships, barges, buoys) in a wide range of depths. It minimizes the time, handling, and equipment required for installation.

Advertised Nominal Holding Capacity

No fixed value. Anchor is usually custom-built, and size is readily varied over a wide range. Usual range is 600 lb to 24,000 lb (weight in air)

Resistance to uplift is approximately 85% of weight in air

Resistance to horizontal force is variable, nominally 20% to 200% of weight in air depending upon seafloor

Nominal Penetration

Hard seafloor:	0 ft
Sediments:	Variable, depending upon soil properties, water depth, anchor size

Water Depth

Design values —
Maximum:	20,000 ft
Minimum:	100 ft (approx, for largest anchor)

Experience —
Maximum:	20,000 ft
Minimum:	50 ft

Limitations

Very heavy anchors in great depths are not retrievable with anchor line

Advantageous Features

Installation time minimized through elimination of on-station ship operations, such as embedment or setting of anchor

Deployable in relatively rough water

3.20.3. Details

Anchor Assembly (see Figures 3.20-1 and 3.20-2)

Drag skirt diameter —
For 18-in.-OD size:	—
For 30-in.-OD size:	—
For 40-in.-OD size:	56 in.

Minimum height of assembly —
For 18-in.-OD size:	2-1/2 ft
For 30-in.-OD size:	3 ft
For 40-in.-OD size:	5 ft

Maximum height of assembly —
For 18-in.-OD size:	4-1/2 ft
For 30-in.-OD size:	6 ft
For 40-in.-OD size:	13 ft

Minimum weight of assembly —
For 18-in.-OD size:	600 lb
For 30-in.-OD size:	1,400 lb
For 40-in.-OD size:	4,000 lb

Maximum weight of assembly —
For 18-in.-OD size:	3,000 lb
For 30-in.-OD size:	6,000 lb
For 40-in.-OD size:	24,000 lb

Nose Cone

Thickness —
For 18-in.-OD size:	—
For 30-in.-OD size:	—
For 40-in.-OD size:	—

Weight —
For 18-in.-OD size:	400 lb
For 30-in.-OD size:	900 lb
For 40-in.-OD size:	2,000 lb

First edition—Blanks will be eliminated as revisions are made.

December 1974

FREE-FALL ANCHOR SYSTEM

Wafers

  Thickness —

For 18-in.-OD size:	—
For 30-in.-OD size:	—
For 40-in.-OD size:	3 in.

  Weight —

For 18-in.-OD size:	200 lb
For 30-in.-OD size:	500 lb
For 40-in.-OD size:	1,000 lb

Cable Pack

  Maximum weight of cable —

Any size:	6,000 lb

Cable (Wire Rope)

  Maximum size —

For 1 x 19 stranding:	3/8 in.
For 3 x 19 stranding:	1-1/8 in.
For 3 x 46 stranding:	1-1/2 in.
For 6 x 19 stranding:	1-1/2 in.

Chain

Length:	25 ft
Size:	5/8 in.

3.20.4. Operational Aspects (see Figures 3.20-3 and 3.20-4)

Operational Modes

  Free-fall installation with cable deployed from cable pack(s) on the anchor

Safety Features

  Accommodates standard field safety practice

3.20.5. Cost

The cost per anchor ranges from $600 for a 600-lb anchor to $30,000 for a $24,000-lb anchor.

3.20.6. References

1. AC Electronics, Defense Research Laboratories. Manual No. OM69-01: Technical manual for Project BOMEX free-fall anchor systems. Santa Barbara, CA, Feb 1969. (Contract no. E-118-69(N)).

2. Delco Electronics. Report No. TR71-05: Containerized cable stowage, by J. Melendez. Santa Barbara, CA, Mar 1971. (Contract no. N00024-70-C-5474)

3. Letter, C. D. Leedham (Delco) to R. J. Taylor (CEL), Sep 20, 1973.

FREE-FALL ANCHOR SYSTEM

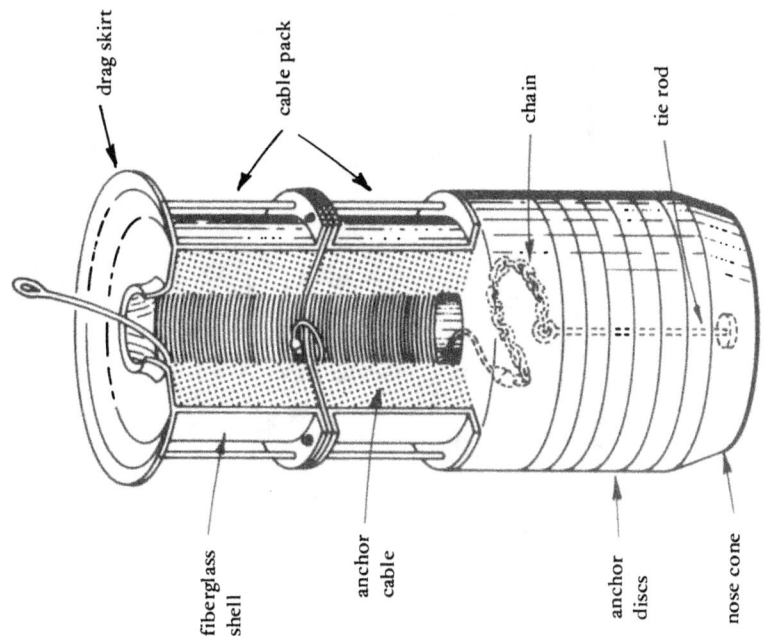

Figure 3.20-2. Delco free-fall anchor (typical).

Figure 3.20-1. Delco free-fall anchor (12,000-pound anchor).

December 1974

FREE-FALL ANCHOR SYSTEM

Figure 3.20-3. Delco free-fall anchor (24,000 pound); mounted on launching platform on USCGS Rockaway.

December 1974

FREE-FALL ANCHOR SYSTEM

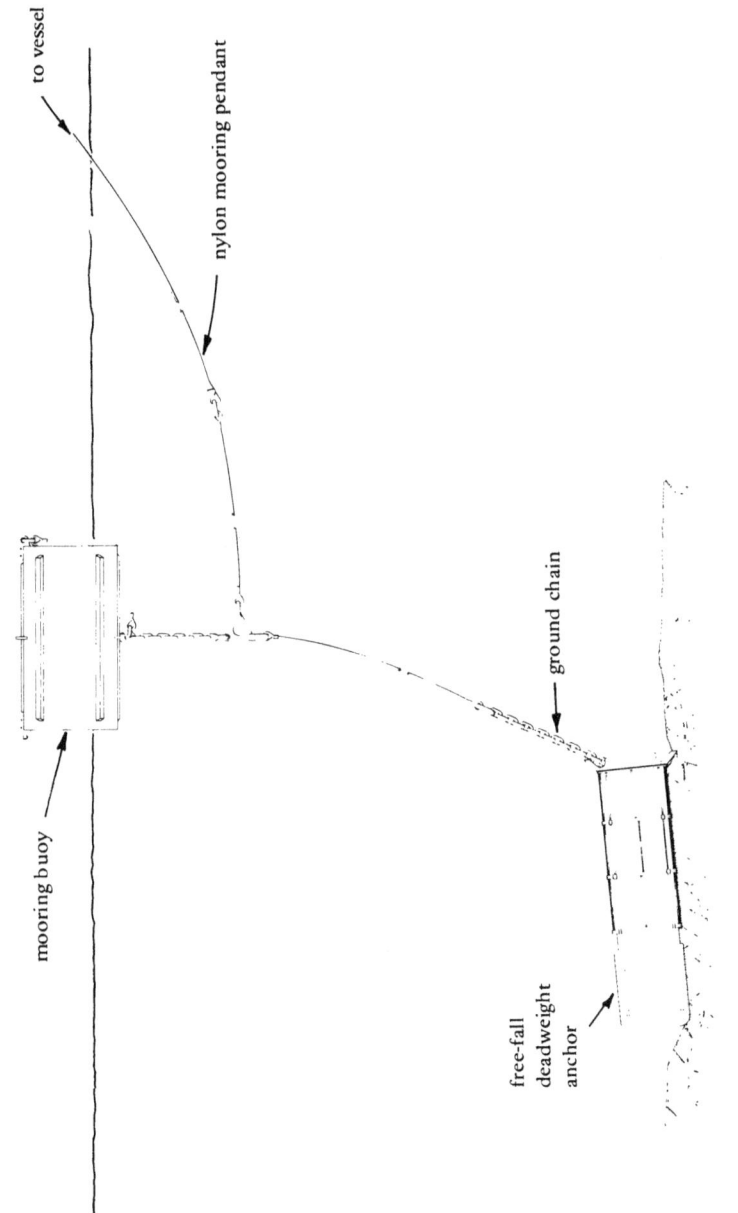

Figure 3.20-4. Delco free-fall anchor in typical deep-water mooring system.

December 1974

Chapter 4. OTHER PROSPECTIVE TYPES

This chapter presents anchors that are still in the conceptual phase or initial phase of development or whose development were terminated due to technical problems. The Implosive, Free-fall, Pulse-Jet, Padlock, Jetted-In and Hydrostatic anchors and Seafloor Rock Fasteners are considered.

4.1. IMPLOSIVE ANCHOR

4.1.1. Background

The implosive anchor concept has only recently evolved. It utilizes hydrostatic pressure as the energy source to embed a projectile into the seafloor. While the idea of the implosive anchor is new, the thought of using the abundant ocean energy to perform useful work is not new. Dantz and Ciani (1967), who were concerned with developing power sources for the deep ocean, designed and built a single-impulse, hydrostatically powered ram device. The usefulness of this power source was verified. Frohlich and McNary (1969) designed and tested a hydrostatically actuated rock corer. They encountered some mechanical problems, but proved that the concept was feasible. The North American Rockwell Corporation actually fabricated an implosive anchor during the 1960s; however, information on this device could not be obtained.

4.1.2. Description

The implosive anchor (Rossfelder and Cheung, 1973) detailed in Figures 4.1-1 and 4.1-2 is similar in form to the propellant-actuated anchor in that it consists of two basic assemblies: the propelled part and the reactive part. The propelled part can either be mounted on an inner piston, which is displaced within a hypobaric breech by admission of the environmental pressure, or it can be the hypobaric chamber itself. The reactive part can be fitted with a shroud to increase its added mass and limit its recoil, or for the case where the chamber itself is propelled, the reactive part can be either an inner piston with shaft and shroud or a free-inertial piston.

4.1.3. Current Status

A feasibility study of the implosive anchor, which included development of a parametric model and performance of a parametric analysis, was conducted (Rossfelder and Cheung, 1973). The major findings were that: (1) anchor operation is influenced by chamber and environmental pressure differential, chamber volume, projectile mass and reactor effective mass, head losses at water entrance, and recoil losses; (2) piston and seals friction appear insignificant for design purposes; (3) for a given anchor mass at a given depth and with a given kinetic energy requirement, there is an optimum volume and geometric design of the hypobaric chamber; and (4) short stroke chambers appear more efficient than long stroke chambers. The study concludes that the concept is feasible and that the primary areas which remain to be addressed are design of the water admission device to minimize head loss, reactor design, and triggering mechanism design.

4.2. FREE-FALL ANCHOR

4.2.1. Background

A "free-fall" anchor is one that falls freely to the seafloor and embeds through its own kinetic energy. Though holding capacities would be limited

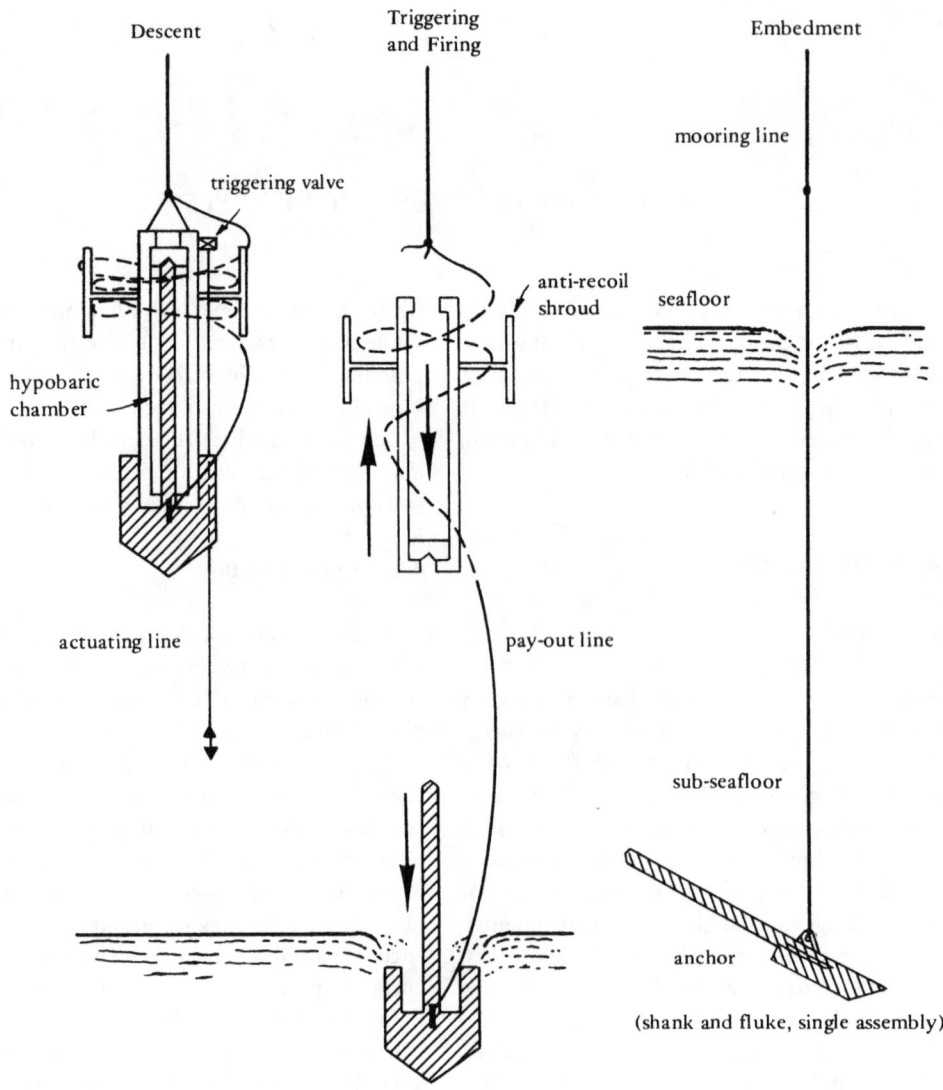

Figure 4.1-1. Propelled-shaft embedment of implosive anchor (Rossfelder and Cheung, 1973).

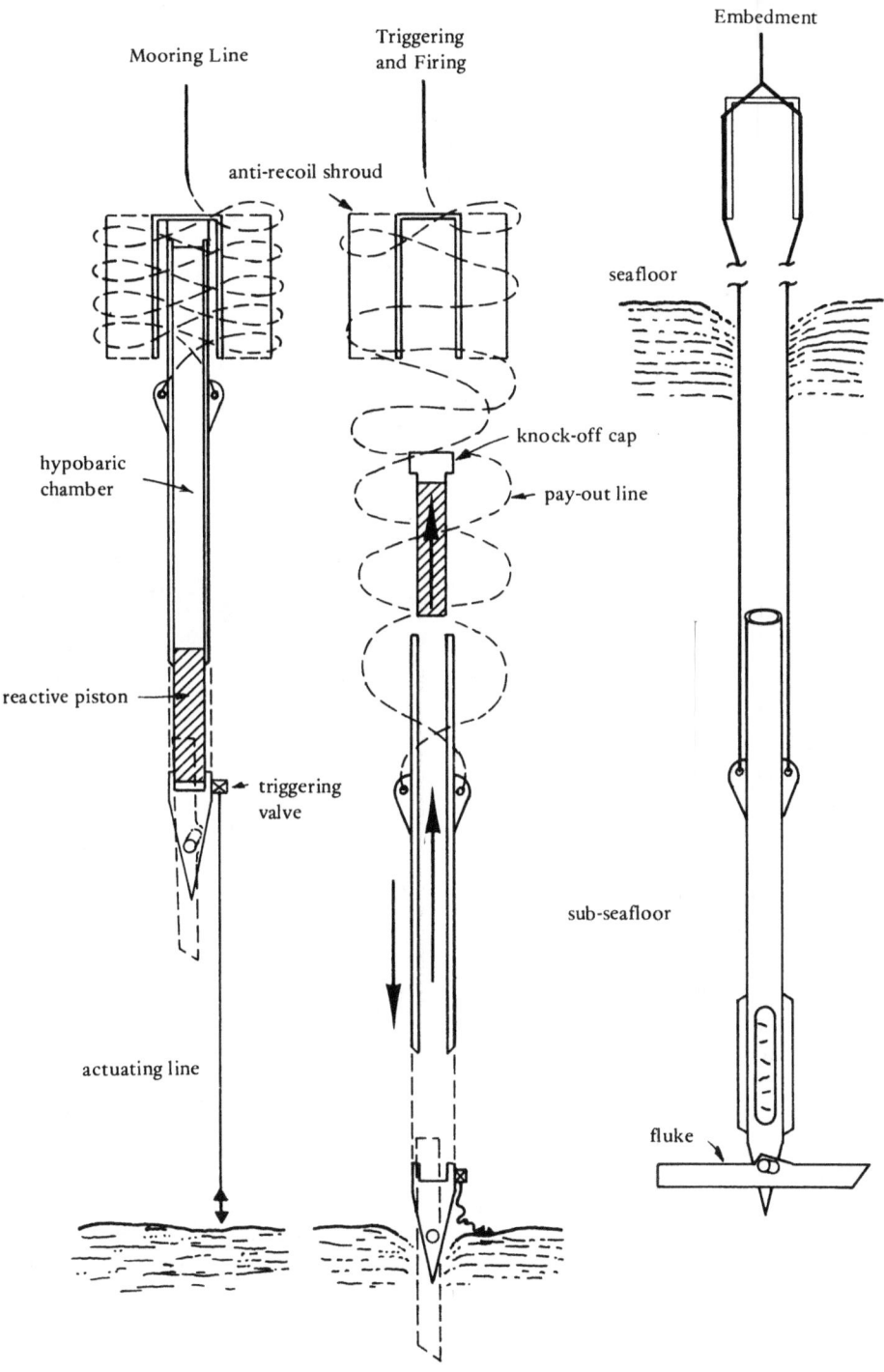

Figure 4.1-2. Propelled-casing embedment of implosive anchor (Rossfelder and Cheung, 1973).

to moderate values, many urgent requirements for anchoring relatively small structures could be satisfied. Quick, easy, and more accurate placement of anchors could be achieved, and better holding power efficiency as measured by holding-power-to-weight ratio could be attained. Holding capacities of 15,000 to 25,000 pounds were considered adequate values to meet these requirements.

4.2.2. Description

After minor modifications to the initial design, the CEL free-fall anchor, Figure 4.2-1, evolved. It is a steel construction in the general shape of an arrow, and it consists of three basic components: a fluke assembly, a heavy steel shank, and a cable bale with protruding fins at the trailing end.

The anchor fluke is a special design which presents a minimum resistance to penetration and keys (rotates from the vertical to horizontal resistive position) rapidly to optimize use of anticipated limited penetration. The cable bale consists of cable coiled in a compact package; a reverse twist is placed in the cable for each coil. Without this reverse twist, the cable would tend to birdcage during pay-out, resulting in greatly reduced line strength and life.

4.2.3. Current Status

As reported by Smith (1966) the free-fall anchor did not fulfill the requirement for being a practical, usable deep-sea anchor that could be free-dropped and, by its own impetus, embed into the seafloor and develop a holding capacity of sufficient amount to warrant its use in place of deadweights. The primary reason was that the size and configuration of the anchor necessary to accommodate the cable bale combined with the size and shape of the flukes necessary to obtain reasonable holding power was not compatible with attaining the velocity needed to obtain adequate embedment. For example, it was determined that even with the maximum theoretical velocity attainable by free-fall (about 35 fps), a holding-capacity-to-weight ratio of only 3 or 4 to 1 could be obtained. A minimum ratio of 7 to 1 is considered necessary for the free-fall anchor to be feasible.

Despite failure to achieve the idealized goal for a free-fall anchor, significant contributions toward development of improved, direct-embedment deep-sea anchors were realized. The cable pay-out system for deploying anchors in the deep sea works and has practical application within certain operational, size, and depth limitations. The knowledge and experience gained can be utilized in deploying future deep-sea anchors. More important is the revolutionary fluke incorporated into the design of the free-fall anchor. This fluke proved to be highly efficient and is adaptable to other types of direct-embedment anchors.

Figure 4.2-1. Free-fall embedment anchor.

4.3. PULSE-JET ANCHOR

4.3.1. Background

The concept for a pulse-jet anchor evolved during the investigation of explosive anchors at CEL. It became evident during testing of the explosive anchors that a power action extending throughout the embedment phase of anchor placement would more readily accommodate the variable resistance to penetration offered by seafloors comprised of firm and soft sediments. The pulse-jet principle could potentially achieve the goal of extending the time during which power is applied to embed the anchor. The concept was investigated under contract by Sea Space Systems, Incorporated. The contractor was to design and fabricate two experimental models and to conduct developmental testing. Then two prototype models were to be delivered for Government testing.

The concept proved to be not feasible, and the contract was reduced in scope to include a report on the effort (Lair, 1967).

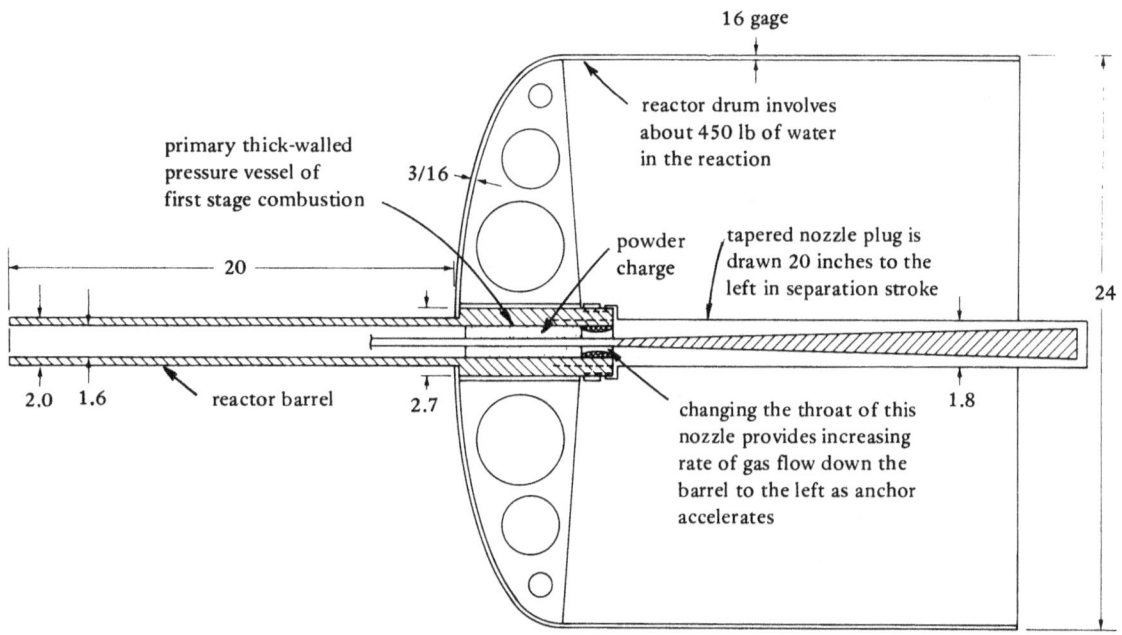

Figure 4.3-1. Mass drag reactor of the Pulse-Jet Anchor System (Lair, 1967).

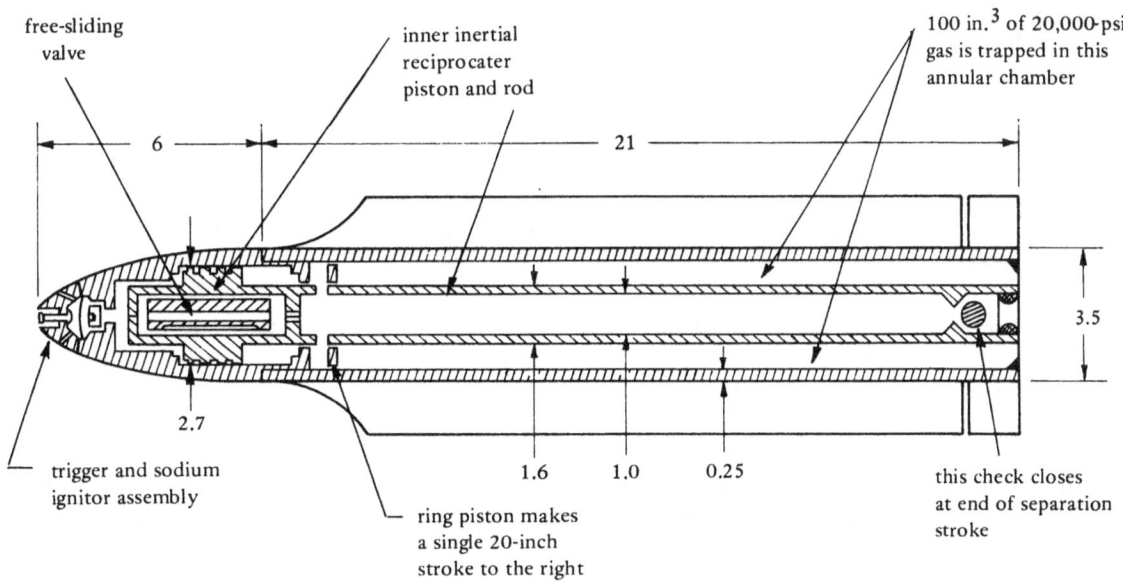

Figure 4.3-2. Ballistic embedding anchor of the Pulse-Jet Anchor System (Lair, 1967).

4.3.2. Description

The pulse-jet anchor as envisaged is comprised of two principal parts: a Mass Drag Reactor, Figure 4.3-1, and a Ballastic Embedding Anchor, Figure 4.3-2. The Ballistic Embedding Anchor is meshed with the Mass Drag Reactor, and the resulting assembly is lowered to the seafloor. On contact, a propellant in the Mass Drag Reactor gives the Ballistic Embedding Anchor an impetus to embed at least its own length into the seafloor. To this point, the principle is similar to that for other propellant-actuated anchors. The Ballistic Embedding Anchor consists of three main components: a main structural body, an inner inertial reciprocator that executes a short stroke with respect to the structural body, and an innermost free-sliding valve that executes a shorter stroke than the reciprocator and governs the stroke of the latter.

As the anchor is expelled from the Mass Drag Reactor, it traps and seals a charge of explusion gases at about 20,000 psi. Beyond this point the principle differs from that of other explosive anchors. This charge of gas is distributed by the valve to drive the reciprocator up and down and ultimately is exhausted forward from the anchor nose to break up the seafloor in front of the advancing anchor. The embedment phase ceases when the gas pressure equals that of the ambient sea. Then a load is applied to the anchor to key it over to a position of maximum resistance.

4.3.3. Current Status

The contractor was unable to achieve an experimental model of the design envisaged. Two ideas were reported as being too optimistic. The first related to the reciprocating machine in that sliding seals could not be made to function satisfactorily at the high temperatures and pressures encountered in the design. The second pertained to determining the critical relationship between the internal mechanics of the anchor and the soil mechanics of the seafloor. Extensive and expensive developmental testing was indicated for both problem areas with no assurance of success.

Two ideas were reported to have stood up under study and evaluation. The first was the concept of multiphase release of energy. The second was the forward jetting of exhaust gases to assist and regulate anchor embedment.

On review of the contractor's report, it was concluded that the cost to solve the problems for successful development of this concept was too great to warrant further investigation.

4.4. PADLOCK ANCHOR SYSTEM

4.4.1. Background

The PADLOCK anchor was designed to provide a high-capacity fixed-point (resistance to bearing, lateral, and uplift loads) anchoring system that could be installed without diver assistance. A feasibility program was initiated at CEL. The scope included the conception, design, fabrication, and evaluation of a self-contained anchor system that employs multiple bearing pads in conjunction with propellant-actuated anchors. The effort, currently suspended, was reported by Dantz (1968).

4.4.2. Description

The PADLOCK is a tripod framework constructed of lightweight materials and supported at each leg by articulated, round bearing pads. To obtain resistance to uplift, propellant-actuated direct-embedment anchors are incorporated into the system. The general scheme of the entire system is shown in Figure 4.4-1. The bearing pads are connected to the frame with ball-joints that allow the pads to maintain maximum contact with the seafloor by adjusting to contour slopes as great as 10%. An anchor is housed above each of the bearings. After the anchors are propelled into the seafloor, they are set by pretensioning the embedment anchor cables with a rewind mechanism located in a central housing unit at the junction of the arms of the tripod framework. The objective is to clamp the pads to the seafloor by obtaining a firm hold in the seafloor soil with the anchors.

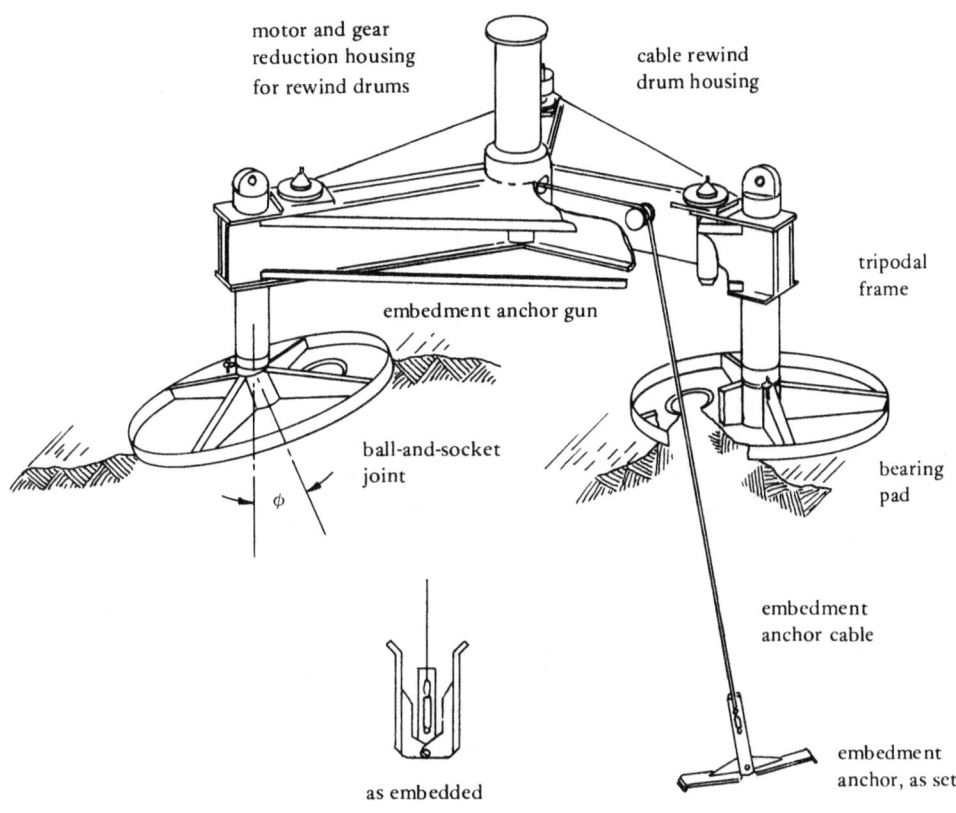

Figure 4.4-1. Basic concept of PADLOCK Anchor System (Dantz, 1968).

A propellant-actuated anchor was selected to develop the uplift resistance. The particular anchor design chosen was the Hove II (now VERTOHOLD) anchor. The commercial anchor of this style was rated as having a nominal 10-kip capacity, whereas a 20-kip capacity was desired. Therefore, the manufacturer had to build and deliver a specially enlarged size. The configuration, size, and load-supporting capacities selected were judged sufficient to demonstrate the feasibility of the system.

The PADLOCK prototype fabricated for testing and evaluation is shown in Figure 4.4-2. A key feature of the concept is the cable rewind mechanism that pulls the anchors to a set position. The rewind mechanism consists of three separate cable drivers powered by a common shaft. Each drum holds the cable from one of the embedment anchors, and each could wind a sufficient length of cable to develop the pretension load for that anchor. Other features of the concept include: (1) an activator unit to control the sequence of operations of the PADLOCK by acoustic command once it is on the seafloor, (2) an ambient-pressure battery power source, and (3) a shipboard stern roller to assist in the installation of the PADLOCK.

4.4.3. Current Status

Five shallow water tests were conducted with the PADLOCK in and about Port Hueneme Harbor in water depths from 18 to 60 feet. The seafloor was primarily hard-packed silty sand. In no single test did all of the components function as a complete system. However, each component performed separately as intended, at least once. Most of the problems involved the propellant-actuated anchors. For

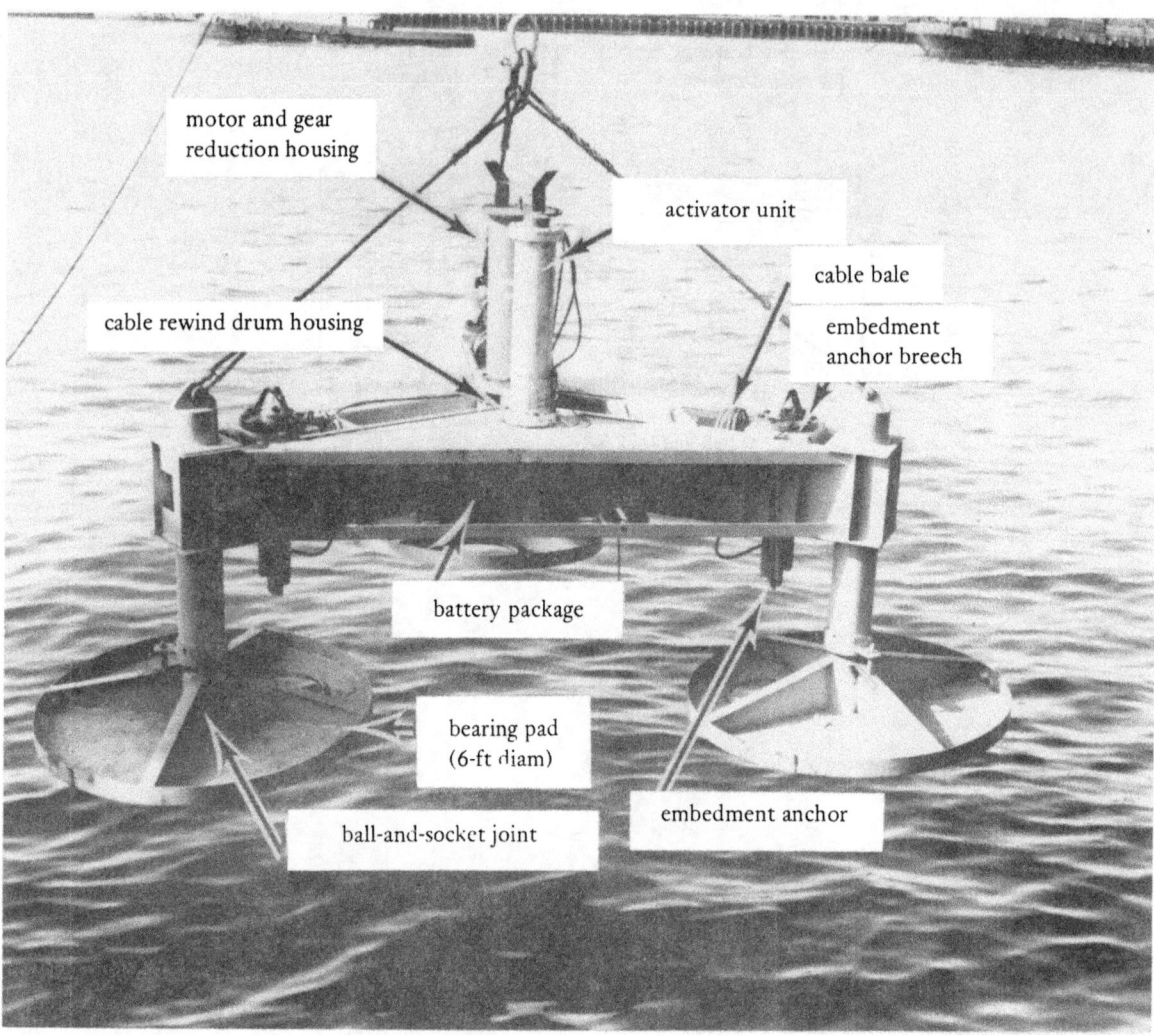

Figure 4.4-2. PADLOCK Anchor System developed for test and evaluation (Dantz, 1968).

example, the contractor-procured anchors were found to be improperly heat-treated, and they failed under high acceleration-induced stress. This fault was corrected after two tests. The recoil of the anchor gun assembly was restricted by the tripod framework, thereby causing high stresses in the anchor and the framework. Problems were encountered with the cable pay-out system. The cable bale had to provide a sufficient amount of cable for the anchor, whose depth of penetration varied for each shot, and a means had to be provided for the rewind mechanism to draw off the remaining cable and develop a pretension in the line. A new frame was designed specifically to accommodate a workable cable pay-out system. The new structure then performed according to design.

The activator unit initially malfunctioned due to an intermittently operating transistor. After the trouble was remedied, the unit functioned according to design. The battery power source was initially used without a protective container (heavy grease provided insulation from seawater), and it was subject to deterioration. Later a battery container filled with transformer oil and covered with a flexible neoprene top to make the system pressure-compensated was used to prevent deterioration of the batteries.

Dantz (1968) concluded that:

"1. In general, the PADLOCK Anchor System has been demonstrated to be a workable concept.

2. The power supply, rewind mechanism, and cable system are workable and fully dependable.

3. The activator unit is operational, water tight at pressures up to 500 psi (no upper limit established), and not affected by the shock loads imposed by the detonation of the embedment anchors.

4. According to a limited number of tests, the reliability of all the components functioning as a complete system is very low, mainly because the reliability of the embedment anchors was unsatisfactory."

In 1968 it was recommended that further effort be suspended until the reliability of propellant-actuated embedment anchors was improved.

4.5. JETTED-IN ANCHOR

4.5.1. Background

The jetted-in approach to anchor embedment can be and/or has been applied to a variety of anchor types, including piles, deadweights, mushroom anchors, and simplified cone anchors. These inexpensive, diver-emplaced jetted-in anchors are capable of sustaining low-to-moderate uplift loads (2 to 10 kips). These anchors would be used for pipe and cable tiedowns, instrument pack tiedowns, and pulling points for underwater construction.

This procedure is considered more applicable in sand seafloors due to the liquefaction potential of this medium. Limited experimental data are available on the increased capacities of large jetted-in anchors; however, there have been tests run on small diver-emplaced anchors that are pertinent to this handbook. This discussion pertains solely to small cone anchors as reported by Stevenson and Venezia (1970).

4.5.2. Description

The jetted-in anchor, Figure 4.5-1, is a buried vertical pipe that is forced into the seabed by a jetting action. Water is pumped into the upper end of the pipe and discharged at the bottom, thereby dislodging soil and permitting the pipe to settle in the hole. An enlarged section, such as a cone-shaped shield at the bottom, and backfilling and grouting the hole are means for improving the holding capacity.

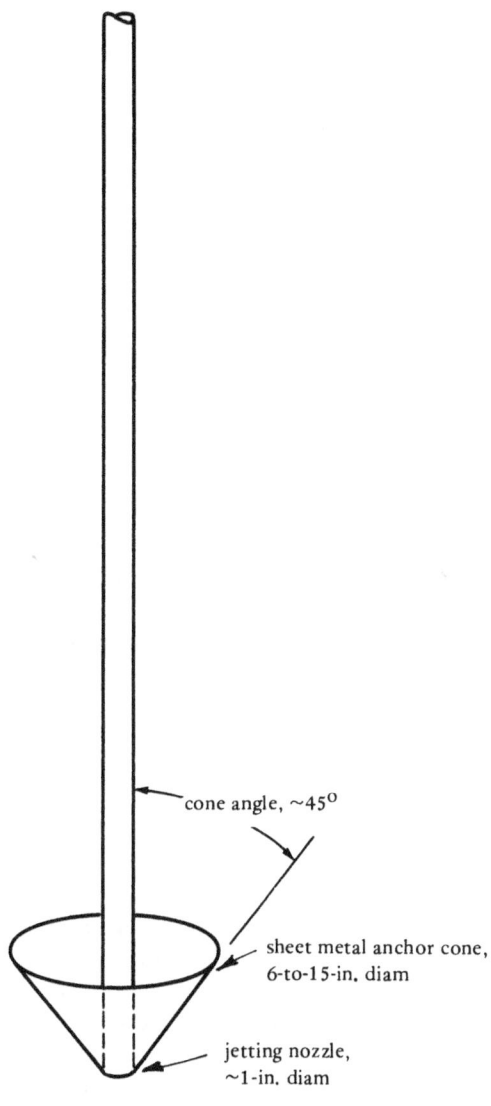

Figure 4.5-1. Illustration of Jetted Anchor.

4.6.2. Description

The hydrostatic anchor, Figure 4.6-1, is comprised of an anchor platform, a penetration skirt, a pump, a lifting harness, and a porous stone. The porous stone is necessary to prevent liquefaction of the soil beneath the stone.

4.6.3. Current Status

According to Wang et al. (1974) the vertical breakout behavior of the hydrostatic anchor depends greatly upon the anchor geometry (including anchor diameter and skirt length), soil strength properties, and the pressure difference between the ambient pressure and the pressure beneath the porous stone. The results of model tests indicate that the hydrostatic anchor functions most effectively in sand with decreasing effectiveness in silts and clays.

4.7. SEAFLOOR ROCK FASTENERS

4.7.1. Background

Seafloor anchors available for shallow-water installation include a variety of seafloor rock fasteners, such as rock bolts, rebar, and drilled and grouted chain. Diver-installed fasteners have been used extensively to stabilize oceanographic cables, to secure structures to rock seafloors, and to moor small vessels. CEL has been attempting to improve the equipment and techniques for installing and, where applicable, grouting the fasteners to the seafloor (Brackett and Parisi, 1975; Parisi and Brackett, 1974).

This section refers specifically to the rock bolt type of seafloor fastener and is generally derived from Brackett and Parisi (1975).

4.7.2. Description

Little data are available on grouted rock bolts; this section will be confined to the nongrouted type. All nongrouted rock bolts utilize the same principle to develop their anchoring strength. By mechanically expanding the down hole end of the bolt, an

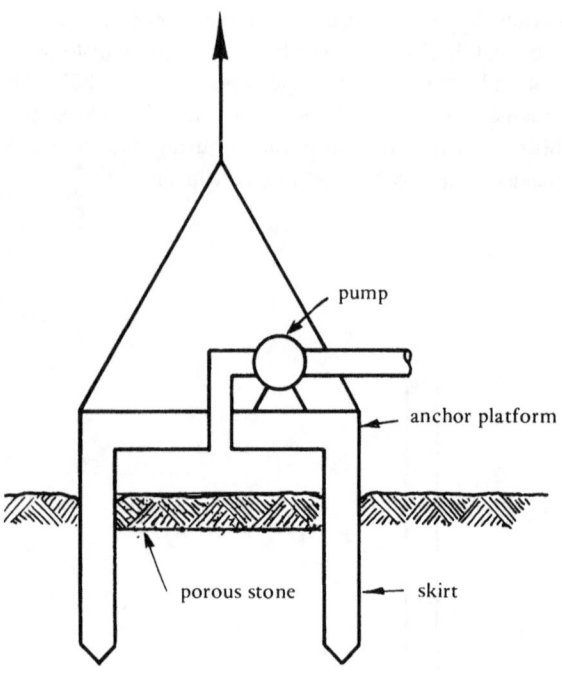

Figure 4.6-1. Schematic of the Hydrostatic Anchor.

4.5.3. Current Status

Twenty-three anchors were diver-emplaced in coral sand; holding capacities varied from 2 to 10 kips. The installation procedures were simple and posed no problems to the divers. The grouting technique was very time consuming and needs refinement.

4.6. HYDROSTATIC ANCHOR

4.6.1. Background

The need for an anchor that could provide short-term vertical resistance to breakout of submersibles and bottom resting platforms was evident. To satisfy this need, work was initiated at the University of Rhode Island on the development of a short-term high-efficiency anchor that utilized suction to develop its capacity (Brown and Nacci, 1971).

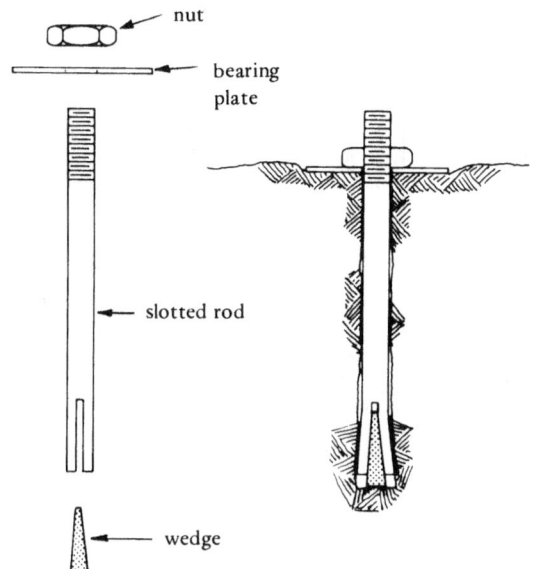

Figure 4.7-1. Drive-set rock bolt, slot and wedge type (Brackett and Parisi, 1975).

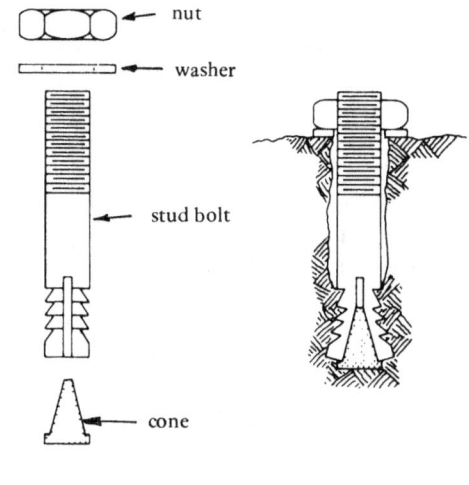

Figure 4.7-2. Drive-set rock bolt, cone and stud anchor type (Brackett and Parisi, 1975).

anchoring force is obtained through a combination of friction, adhesion between the anchor and rock, and physical penetration of the anchor into the rock. Rock bolts can generally be classified into two types: (1) drive-set, and (2) torque-set.

The slot and wedge bolt, Figure 4.7-1, and cone and stud anchor, Figure 4.7-2, are common examples of the drive-set type. The anchor is secured by placing the wedge into the slot and positioning the rod into the predrilled hole, then by driving the slotted rod over the wedge (which rests on the bottom of the hole) the rod expands into the rock.

Successful installation of the drive-set fastener depends on accurate hole drilling to a predetermined depth and the application of sufficient force to completely expand the slotted rod. Problems can also be encountered in soft rock where the driving force causes the wedge to be pushed into the rock rather than expanding the anchor.

A typical torque-set anchor is shown in Figure 4.7-3. This type of rock anchor has a wedge or cone that is threaded to the bottom of the bolt. A sleeve or shell that surrounds the cone is pushed into the hole with the bolt. Once the bolt has been inserted, torque is applied to the nut to pull the bolt and cone up through the sleeve, thus securing the anchor.

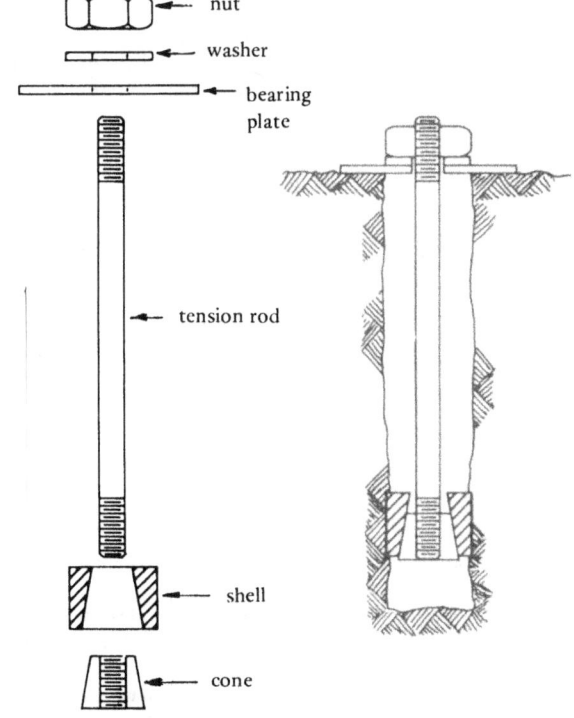

Figure 4.7-3. Torque-set rock bolt (typical) (Brackett and Parisi, 1975).

Table 4.7-1. Parameters Affecting Holding Strength of Seafloor Fasteners

Parameter	Effect on Holding Strength	Comments
Bolt diameter	The bolt diameter determines the ultimate potential holding strength possible for a given size bolt, and the ultimate tensile strength	If all bolts have the same ultimate tensile strength, the failure load of the bolt will vary as the square of the diameter.
Anchor configuration		
Length and diameter of collar	The length and diameter of the anchor collar affect the stress produced in the seafloor rock. An increase in size of the anchor collar will decrease the stresses in the rock, thus reducing the chance of failure due to localized crushing or splitting of the rock.	An increase in anchor diameter requires an increase in drilling time. The trade-off between installing one large rock bolt or several small bolts in a padeye configuration should be considered.
Type of collar	A one-piece split collar has proven to give slightly higher pullout loads than the two-piece collar design for the same size fastener.	
Embedment depth	An increase in embedment depth produces almost a linear increase in holding strength up to the point where either localized crushing of the rock occurs around the collar or the ultimate tensile strength of the bolt is exceeded.	As a general rule a 6-inch embedment is sufficient to eliminate failure due to surface fracturing of the rock. Bolt diameter, competency of the rock, and presence of hard or soft substrata should be considered before determining the minimum embedment depth.
Duration of installation	There is not sufficient data at the present time to predict the exact effect of corrosion on the long-term holding strength of the fasteners tested. A trend toward a slightly reduced holding strength was detected after as little as 6 months of exposure.	The use of zinc anodes along with periodic inspection and replacement of spent anodes should ensure the integrity of the fastener for many years.
Initial torque	Initial torques of 40 ft-lb for the masonry stud anchor and 100 ft-lb for the spin-lock rock bolt were found to be necessary to properly set the anchor. Torquing the bolts above these values have no effect on the holding strength of the bolt.	The masonry stud anchors could be properly set by a diver using a hand wrench, but the use of an hydraulic impact wrench is recommended to ensure proper setting of the spin-lock rock bolt.
Compressive strength of rock	The holding strength of a given size fastener is almost linearly dependent on the unconfined compressive strength of the rock.	The presence of internal voids or fractures in the rock must be investigated before using compressive strength as a design criterion.
Installation of fasteners on land versus underwater	There appears to be a slight decrease in holding strength for bolts installed underwater compared to the same installation on land. The wide scatter of data points makes it difficult to quantitatively determine the magnitude of this decrease in failure load. However, if a normal safety factor is applied to the results of land tests, a realistic safe working load for the underwater installation should be obtained.	Care must be taken when using land tests to predict underwater performance. The test installations must be conducted in rock representative of that actually found at the seafloor work site. This analysis should include: size, porosity, presence of voids and fractures, presence of biological organisms, such as those in coral, that may have a significant effect of the holding strength of the fastener.

The torque-set bolt requires far less precision in hole drilling providing the depth is greater than the length of the bolt. Expansion of the anchor is also unaffected by the quality of the rock at the bottom of the hole.

With the hand-held and hydraulically powered tools currently available to the underwater construction and salvage divers, it is easier to provide the torque for installing the torque-set type of anchor than the linear impact for installing the drive-set type.

4.7.3. Current Status

Table 4.7-1 summarizes the parameters affecting the performance of seafloor rock bolts.

Work to date on diver-installed grouted fasteners has primarily involved development of a grout-dispensing device. The device is workable but must be lightened prior to Fleet usage. Testing on grouted fasteners has been minimal, but results indicate that the rock bolt type of fastener is superior to grouted fasteners because it is far simpler and quicker to install.

Chapter 5. APPLICABLE COMPUTATIONS

The determination of the holding capacity of anchors designed to resist uplift loads involves considerations and techniques not required for conventional anchors. Conventional anchors are designed to embed as they are dragged. Should applied loads exceed their capacity, they will displace laterally but generally will continue to maintain their approximate design holding capacity once the excess loading eases. However, uplift-resisting anchors must be embedded by some means other than the service loading force, such as by drilling, driving, or ballistic propulsion. Once the uplift-resisting anchor is at its deepest penetration achieved during installation, all subsequent in-service applied loads will tend to extract it. Slight initial upward movements tend to seat it and mobilize the surrounding soil medium to resist extraction. Any excess loading on and/or movement of the anchor causes a reduction of the rated capacity and eventually causes extraction.

It is evident then that determining the penetration of an uplift-resisting anchor is important. Also, determining the initial movements to mobilize the soil and, in the case of anchors with outward folding flukes, determining the fluke-keying distance are important. Thus, in the next section computations to determine penetrations and fluke-keying distances are considered. Then in the following section methods for predicting holding capacities are presented.

5.1. PENETRATION

Penetration depths cannot be analytically predicted reliably in coral and rock. The soils are separated into two categories: clay and sand. Analytical techniques are provided for estimating the penetration of anchors driven ballistically and by vibration.

5.1.1. Momentum Penetration

Momentum penetration is defined as the penetration achieved from its own momentum. The momentum can result from the anchor being fired from a gun, in which case the fluke or projectile is traveling at a high velocity when it strikes the seafloor. Or it can result from its own free-fall impetus. Propellant-actuated, implosive, and free-fall embedment anchors fall in this penetration category.

5.1.1.1. Clay. Momentum penetration in clay can be estimated by the methods established by True (to be published). Equations for the solution of penetration problems are not suitable for a closed-form solution. However, they can readily be solved by incremental techniques. The incremental form to be used for computations is:

$$v_{i+1} = v_{i-1} + \frac{W - F_i(v_i, z_i)}{M^* v_i} (2 \Delta z) \quad (5-1)$$

where
- v_{i+1} = velocity at the depth being considered (ft/sec)
- v_{i-1} = velocity at two depth measurements above the depth being considered (ft/sec)
- W = buoyant weight of projectile in soil (lb)
- v_i = velocity calculated one depth increment above the depth being considered (ft/sec)
- $F_i(v_i, z_i)$ = resisting force at the depth and velocity one depth increment above the depth being considered
  $= F_i^* + F_{H_i}$ (lb)

Table 5.1-1. Values of Side Adhesion Factor, δ^*, at High Velocity Derived From Field Test Data

Projectile Shape	Slenderness Ratio, ℓ/D	High-Velocity Side Adhesion Factor, δ^*
Stubby	9	0.11
Medium	15	0.23
Slender	30	0.46

M^* = effective mass of penetrator; equals penetrator mass plus added mass (slug)

Δz = depth increment (ft)

F_i^* = soil resisting force = $C_{1_i} S_{u_i} S_{\dot{e}_i}$ (lb)

F_{H_i} = fluid inertial drag force = $v_i^2 C_2$ (lb)

S_{u_i} = undrained sediment shear strength (psf)

$S_{\dot{e}_i}$ = ratio between dynamic and static shear strength
= $S_{\dot{e}_i}^*/1 + \left[1/\sqrt{(C_{\dot{e}} v_i/S_{u_i} l_i) + C_o} \right]$

C_{1_i} = $N_c A_{F_i} + (\delta_i^*/S_{t_i}) A_{s_i}$ (ft²)

C_{2_i} = $(1/2)\rho_i C_D A_{F_i}$

$S_{\dot{e}}^*$ = maximum $S_{\dot{e}_i}$ at high velocity; equal to 5 for all soils

$C_{\dot{e}}$ = constant; equal to 20 for all clays and sands (psf-sec)

l_i = effective length of shearing zone; equals depth of embedment or length of penetrometer body, whichever is smaller (ft)

C_o = dimensionless constant; equal to 0.04 for all clays and sands

N_c = deep bearing factor; equal to 9 for clays and sands

A_{F_i} = frontal area of penetrometer (ft²)

δ_i^* = adhesion reduction factor (see Table 5.1-1)

S_{t_i} = soil sensitivity (ratio of remolded to undisturbed strength); use $S_{t_i} = 1$ for sands

A_{s_i} = side area of penetrometer (ft²)

ρ_i = mass density of soil (slug/ft³)

C_D = drag coefficient (estimated from fluid mechanics principles)

In Equation 5-1, all functions are known except v_{i+1}; Δz is specified at one-twentieth or less of an estimated embedment depth. When beginning, however, $v_i = v_1$ is not known, and it is necessary to estimate v_1; this is done most directly by computing v_2 for $v_1 = v_o$ and then starting over again using

$$v_1 = \frac{v_o + v_2}{2}$$

An equivalent direct relationship for this procedure is

$$v_1 = v_o - \frac{\Delta z}{v_o M^*} \left(C_{2_1} v_o^2 + C_{1_1} S_{\dot{e}_1} S_{u_1} - W \right) \quad (5\text{-}2)$$

A better estimate of an initial v_1 will not give a better value of final depth, z_n. A flow diagram of the calculation procedure is shown in Figure 5.1-1.

5.1.1.2. Sand. Momentum penetration in sand can be estimated with the same techniques and equations used for estimating momentum penetration in clay.

5.1.2. Vibratory Penetration

Vibratory penetration is defined as penetration gained by transmitting high-frequency vibration to an anchor so that under its own and/or additional weight it will sink into the seafloor.

Figure 5.1-1. Incremental calculation flow for momentum penetration in clay and sand.

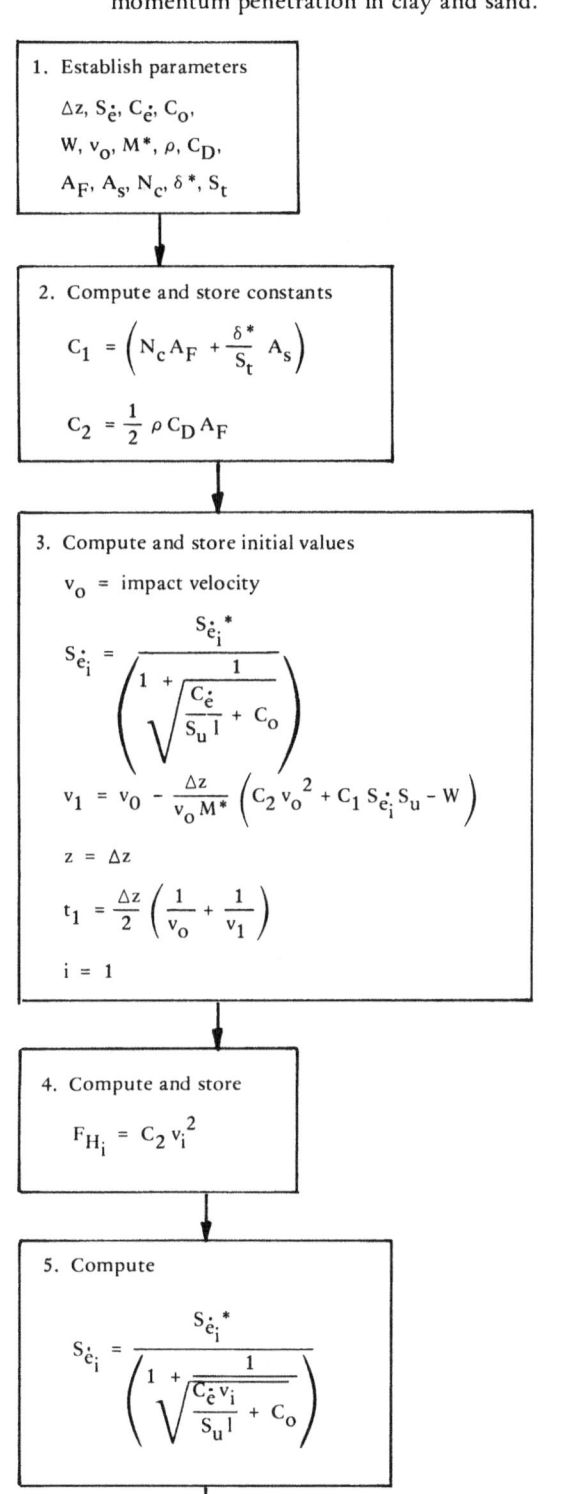
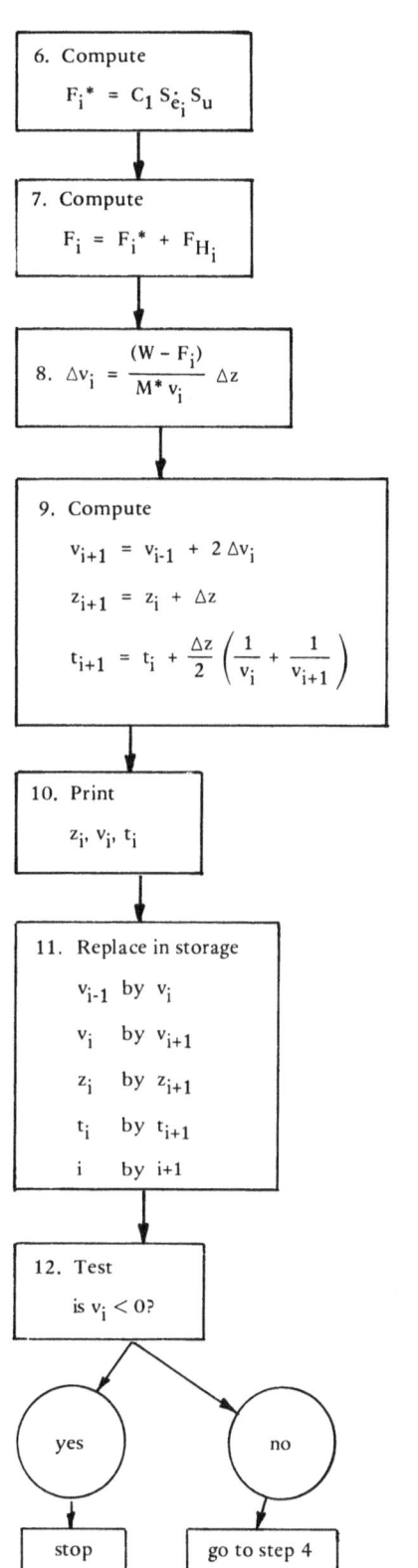

Schmid (1969), who has discussed vibratory penetration in sand and clay soil, states that a vibratory driver will fail to advance the driven object when the total weight of bias plus the peak driving force is about equal to the total soil resistance to penetration. Beard (1973) presented Schmid's equations as applied to anchors with flukes on the lower end of a long shaft to be driven into the seafloor. These equations are presented here.

Clay. Vibratory penetration in clay can be readily calculated with the following equation:

$$Q + Bias = A_{fs} c + A_{ff} N_c c + a_s c_r D \quad (5\text{-}3)$$

where
- Q = peak vibrator driving force (lb)
- $Bias$ = weight of fluke-shaft vibrator system (lb)
- A_{fs} = fluke side area (ft^2)
- A_{ff} = fluke frontal area (ft^2)
- c = soil cohesion (psf)
- N_c = deep bearing capacity factor for clay; equal to 9
- a_s = shaft unit area (ft^2/ft)
- c_r = remolded soil cohesion (psf)
- D = fluke embedment depth (ft)

For clays that have a uniform cohesion profile with depth, the above equation can be solved directly for the embedment depth. When the cohesion profile varies as a complex function of depth, it is necessary to solve the equation by trial and error because a particular cohesion value implies a particular depth. However, for seafloor soils the cohesion profile is often specified by a constant function of depth in the form of a ratio of cohesion to effective overburden pressure. Multiplying this ratio by depth and buoyant soil density gives the cohesion at that depth. (The remolded cohesion is attained by dividing the cohesion by the soil sensitivity.) When this is the case, Equation 5-3 becomes

$$Q + Bias = A_{fs} \frac{c}{p} \gamma_b D + A_{ff} N_c \frac{c}{p} \gamma_b D + a_s \left(\frac{1}{S_t}\right)\left(\frac{c}{p}\right)\gamma_b \frac{D^2}{2} \quad (5\text{-}4)$$

This equation can be solved for depth in terms of the other parameters using the quadratic equation. The result is:

$$D = \frac{-(X+Y) \pm [(X+Y)^2 + 4W(Q+Bias)]^{1/2}}{2W} \quad (5\text{-}5)$$

where
- $X = A_{fs}(c/p)\gamma_b$
- $Y = A_{ff} N_c (c/p)\gamma_b$
- $W = (1/2) a_s (1/S_t)(c/p)\gamma_b/2$
- c/p = ratio of cohesion to effective overburden pressure
- γ_b = buoyant unit weight of soil (pcf)
- S_t = soil sensitivity

Sand. For sand the equation for vibratory penetration is

$$Q + Bias = A_{fs} \sigma_v K \tan\phi_s + A_{ff} N_q \sigma_v + a_s \sigma_v K \tan\phi_s \frac{D}{2} \quad (5\text{-}6)$$

where
- σ_v = effective vertical pressure (psf)
- K = ratio of principal soil stresses
- ϕ_s = friction angle between object and sand (deg)
- N_q = deep bearing capacity factor for sand

It is recommended that N_q values be chosen according to the curve in Figure 5.1-2. Values of K can be taken as 1.5 for dense sand and 1.0 for loose sand. The angle of friction between sand and smooth metal

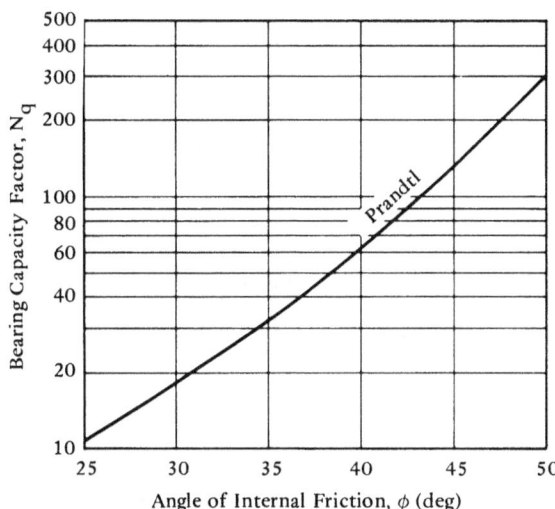

Figure 5.1-2. Theoretical bearing capacity factor, N_q, versus angle of internal friction, ϕ, for a strip foundation.

Table 5.1-2. Ratio of Keying Distance* to Fluke Length

Type of Fluke	Ratio of Keying Distance to Fluke Length
Expandable (finger-like flukes	2 - 3
Rotating Plate fluke	2 - 3
Screw-In	0
Eccentric-keying flat-plate fluke	1 - 2

* Distance measured vertically from fluke tip.

surfaces is independent of soil density and is taken as 26 degrees. For rough surfaces ϕ_s should be taken as the angle of internal friction of the sand. When the density of the sand varies significantly with depth, Equation 5-6 must be solved by trial and error. If the sand has a uniform density over the depth of interest or if it can be approximated as such, Equation 5-6 can be rewritten by substituting the product of soil depth and soil buoyant density for the effective vertical pressure. Equation 5-6 then becomes:

$$Q + \text{Bias} = A_{fs} \gamma_b D K \tan\phi_s + A_{ff} N_q \gamma_b D + a_s \gamma_b D K \tan\phi_s \frac{D}{2} \quad (5\text{-}7)$$

This equation can be solved for depth in terms of the other parameters using the quadratic equation. The result is:

$$D = \frac{-(I+L) \pm [(I+L)^2 + 4J(Q+\text{Bias})]^{1/2}}{2J} \quad (5\text{-}8)$$

where $\quad I = A_{ff} N_q \gamma_b$

$\quad L = A_{fs} \gamma_b K \tan\phi_s$

$\quad J = (1/2) a_s \gamma_b K \tan\phi_s$

5.1.3. Screw or Auger Penetration

Penetration of screw or auger types of anchors can be estimated best by reviewing penetrations achieved in various types of soil.

5.1.4. Penetration Reduction Due to Fluke Keying

The depth of embedment to be used in a holding capacity calculation is not the penetration depth; it is the penetration depth less the distance required to bring the fluke to fluke length for a variety of fluke types. Multiplying these factors by the fluke length will give an estimate of the distance required to key a fluke. These distances are given in Table 5.1-2.

5.2. HOLDING CAPACITY

The purpose here is to provide methods for estimating the holding capacity of uplift-resisting anchors in seafloor soils. Holding capacity cannot be estimated analytically in rock and coral. In those materials field tests and general experience must be relied upon.

5.2.1. Basic Holding Capacity Equation

The maximum uplift forces that can be applied to direct-embedment anchors without causing the anchors to pull out are identified as the anchor holding capacities. Holding capacity is not a property of a particular anchor, but varies considerably with seafloor type, embedment depth, and method of loading.

It is necessary to subdivide the holding capacity problem into categories. The first subdivision is based on general soil type, of which there are two: cohesive and cohesionless. Cohesive soils are fine-grained plastic materials (clays), and cohesionless soil are coarse-grained nonplastic materials (sands). The second subdivision is based on method of loading. For each general soil type three methods of loading will be considered: short-term static, long-term static, and long-term repeated. Short-term static loading describes the situation in which the anchor is loaded rapidly until breakout occurs. Most field tests have been conducted in this manner, and most of the theoretical results are directed toward it. Long-term holding capacities are usually presented as fractions of the immediate capacity. Long-term static holding capacity refers to the situation in which an anchor pulls out after a constant upward force has been applied over a long period of time. This holding capacity would be associated with moored objects such as submerged buoys. Repeated loading involves a line force that varies considerably with time; it can be approximated by a sinusoidally varying force with a certain period and amplitude. Moored surface buoys and ships can provide this type of force application.

The holding capacity problem has been divided into six categories; they are:

1. Cohesive soil — short-term static loading
2. Cohesive soil — long-term repeated loading
3. Cohesive soil — long-term static loading
4. Cohesionless soil — short-term static loading
5. Cohesionless soil — long-term repeated loading
6. Cohesionless soil — long-term static loading

The commonly used equation for representing the holding capacities of embedment anchors is:

$$F_T = A(c\overline{N}_c + \gamma_b D \overline{N}_q)(0.84 + 0.16 B/L) \quad (5\text{-}9)$$

where
A = fluke area (ft^2)
c = soil cohesion (psf), characteristic strength
γ_b = buoyant unit weight of soil (pcf)
D = fluke embedment depth (ft)
$\overline{N}_c, \overline{N}_q$ = holding capacity factors
B = fluke diameter or width (ft)
L = fluke length (ft)

The equation is relatively general and can be applied to almost any form of loading. However, the holding capacity factors and the cohesion may vary with the loading mode, and they have been found to vary with soil type, density, and relative anchor embedment depth, D/B (B is the fluke width). The major problem of estimating holding capacity is then one of estimating c, $\overline{N}_c$, and $\overline{N}_q$.

5.2.2. Holding Capacity Prediction Procedure

The general procedural framework presented here is shown by the block diagram of Figure 5.2-1; each item of the diagram is discussed briefly below. The numbering system below compares with that of the diagram.

In virtually all cases, an anchor should be installed so as to display "deep" behavior. In all curves of holding capacity or holding-capacity-parameters-versus-depth, there are breaks below which the holding capacity increases less rapidly with increasing depth; this behavior in the lower sections of these plots is termed "deep." It is advantageous to establish a "deep" anchor, because errors in locating the anchor, either during installation or because of deformations after installation, do not cause large changes in holding capacity. The anchor is, therefore, more reliable.

A step by step approach for calculating anchor holding capacity is as follows:

(1) Determine Design Parameters. Determine the anchor fluke embedment depth, D (using techniques of section 5.1), width, B, length, L, and projected area, A.

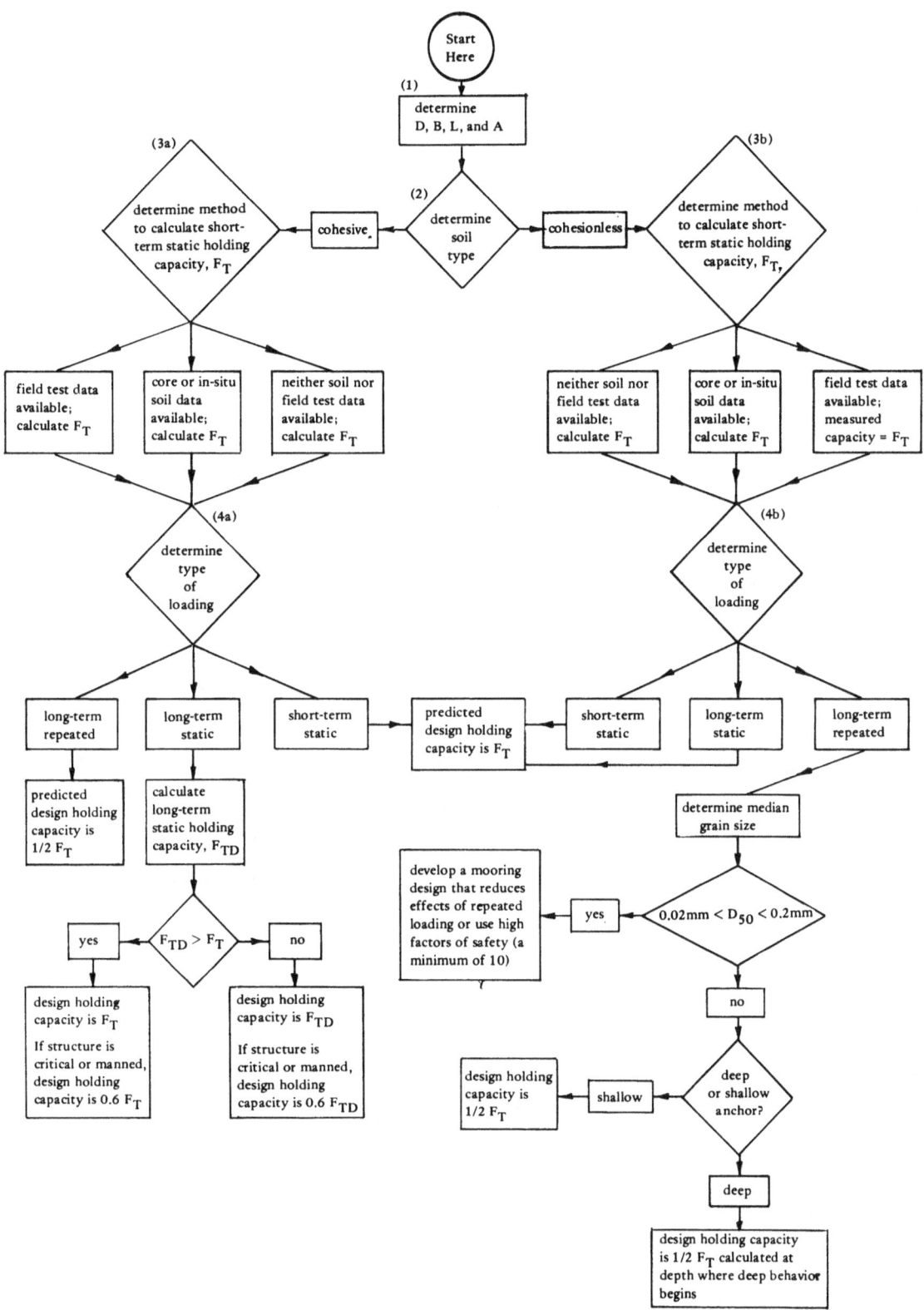

Figure 5.2-1. Prediction procedure for direct-embedment anchor holding capacity.

(2) Determine Soil Type. Determine the general soil type (cohesive or cohesionless). This will be obvious from the visual observation of a bottom sample, even from a very disturbed grab-type sample. In areas far from shore, it may be possible to estimate the type of bottom from a chart of the regional geology. In addition, good geophysical data, if available, may give clues. If at all possible, however, a bottom sample should be obtained.

(3a) Determine Calculation Method for Cohesive Soil. The short-term static holding capacity for *cohesive soils* can be estimated three different ways depending on the data that are available. One way is based on anchor field test data, the second way on good quality core data or in-situ strength data, and the third way is for when no soil data or anchor field test data are available.

Anchor field test data available. Field tests provide a good means for estimating short-term holding capacities. However, field tests in cohesive soils develop the strength of the soil under the anchor (suction forces) and, therefore, need to be modified to account for these suction forces. If this is not done, unconservative design values will result. Figure 5.2-2 can be used to account for the suction effect. Using the relative embedment depth ratio, D/B, and an estimate of c (1 psi should be a reasonable value in most cases), a reduction factor, R, is obtained. This is inserted into the equation given on the figure, and the design short-term holding capacity, F_T, is calculated. An estimate of the soil unit weight, γ_b, is needed and can be assumed to be equal to 25 pcf in most cases.

Core or in-situ soil data available. When core or in-situ soil data are available, the short-term static holding capacity can be calculated from Equation 5-9. Some of the values for this equation must be evaluated. Start by making plots of the undrained or vane shear strength and unit weight distributions. If the strength and density are approximately uniform with depth, then the characteristic strength, c, and density, γ_b, are simply the mean values over the depth range, D. If the strength increases approximately linearly with depth from a value of near zero at the seafloor surface, then the plots of Figure 5.2-3 are used to obtain the characteristic strength and density. This is done by first calculating D/B and taking the strength, c_a, at depth, D (D is the anchor depth after setting), from the strength profile. Figure 5.2-3 is entered with these values, and the quantity D_c/B is determined. D_c/B is the ratio of the distance above the fluke at which the characteristic strength is measured to the fluke width or diameter. The characteristic strength, c, and density are then taken as the strength and density a distance D_c above the anchor fluke. For more unusual strength and density profiles, either a conservative uniform or linearly increasing curve should be drawn through the data, or an experienced seafloor soils engineer should be consulted.

Now that D/B and c are known, the parameter $\overline{N}_c$ can be obtained from Figure 5.2-4. $\overline{N}_q$ for cohesive soils is 1. Now that all the values of the parameters have been determined, the short-term holding capacity, F_T, can be calculated from Equation 5-9 or from the nomographs, Figures C-1, C-2, and C-3, in Appendix C.

Neither soil nor anchor field test data available. When no data are available, soil properties must be assumed to estimate holding capacity. The shear strength and unit weight distributions of Figure 5.2-5 should be used, and the above steps followed to accomplish this. The procedure can be simplified by using Figures B-1, B-2, and B-3 in Appendix B where holding-capacities-versus-depth have been plotted for the operative anchors presented in this handbook. If at all possible, however, strengths and densities for the design locations should be measured, and the steps in the above paragraph followed.

(4a) Determine Type of Loading for Cohesive Soil. Most anchor trial tests, salvage work, and other projects that require a reaction force for a short period of time are considered to be short-term static loadings. Surface vessels and buoys generally exert a long-term repeated loading condition, although certain designs may convert the repeated load into a virtual long-term static condition. Subsurface buoys, suspended arrays, and other suspended structures exert long-term static loads.

Short-term static loading. If the loading is short-term static, the design holding capacity is F_T as determined by the selected method in paragraph 3a above.

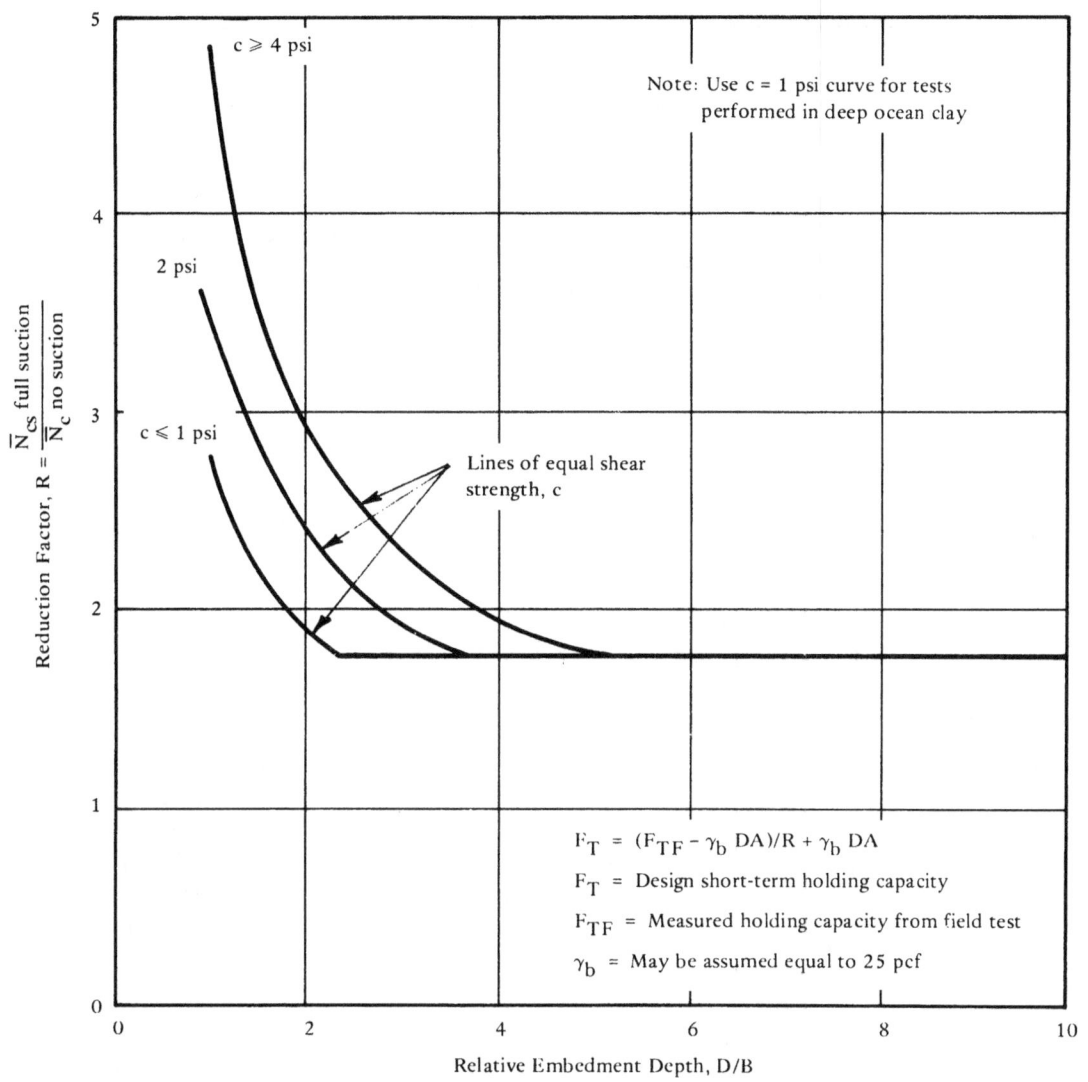

Figure 5.2-2. Reduction factor to be applied to field anchor tests in cohesive soils to account for suction effects.

Long-term repeated loading. If the loading is long-term repeated, the design holding capacity is one-half F_T as determined by the selected method of paragraph 3a above. This capacity refers to the characteristic peak repeated load. The rationale for this reduction has been given by Taylor and Lee (1972).

Long-term static loading. If the loading is long-term static, the long-term capacity, F_{TC}, must be estimated. To do this, parameters for the equation must be evaluated. First, the drained friction angle, ϕ, the quantity D/B, and the parameter $\overline{N}_q$ are obtained.

$\overline{N}_c$ and c are set equal to 0 for long-term conditions. Next, the drained holding capacity, F_{TD}, is obtained from Equation 5-9 (substituting F_{TD} for F_T). F_{TD} is compared with F_T from paragraph 3a above, and the lower value is used as a design holding capacity. If the anchored system is critical or manned, the result should be multiplied by 0.6 to account for possible creep effects. This reduction for creep effects has been explained by Taylor and Lee (1972).

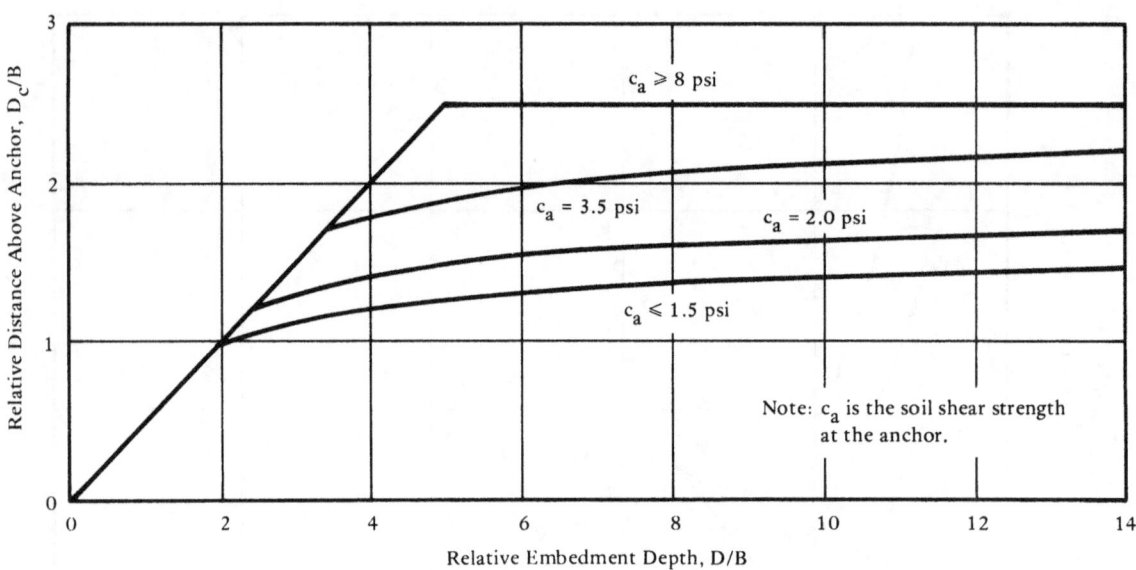

Figure 5.2-3. Plot for calculating D_c, the distance above the anchor, at which the characteristic strength, C_a, is to be taken.

(3b) Determine Calculation Method for Cohesionless Soils. The procedure to be followed in estimating the short-term static holding capacity in *cohesionless soils* depends upon the type of data available. Anchor field test data, core or in-situ soil data, and a lack of data present three approaches for making the required estimate.

Field test data available. The measured holding capacity from a field test can be considered to represent the proper short-term holding capacity, because suction will not be significant in cohesionless soil.

Core or in-situ soil data available. When core or in-situ data are available, Equation 5-9 can be used for estimating the short-term static holding capacity. Values for the parameters in this equation need to be evaluated first. The friction angle, ϕ, and the unit weight, γ_b, in the vicinity of the anchor fluke should be estimated. The parameter $\overline{N}_q$ can be obtained from Figure 5.2-6, given ϕ and D/B. $\overline{N}_c$ and c are equal to 0 in a cohesionless soil. The short-term static holding capacity, F_T, is now obtained from Equation 5-9 or by using the nomographs, Figures C-4, C-5, or C-6 in Appendix C.

Soil or field test data not available. When no data are available, assume the friction angle to be 30 degrees and the unit weight to be equal to 60 pcf. The procedures of the preceding paragraph can be used with these soil properties to determine F_T by Equation 5-9. The procedure can be simplified by using Figure B-4, B-5, or B-6 in Appendix B where holding-capacities-versus-depth for these soil properties have been plotted for the operative anchors presented in this handbook.

(4b) Determine Type of Loading for Cohesionless Soil. The type of loading should be determined in a manner identical to that of paragraph (4a).

Short-term static loading. If the loading is short-term static, the holding capacity is F_T as determined by the selected method in (4a) above.

Long-term repeated loading. If the loading is long-term repeated, the grain size distribution and the relative embedment depth need to be considered. Therefore, a grain size analysis of a soil sample should be performed. If the median grain size (D_{50}) is found to lie between 0.02 and 0.2 mm, either a different

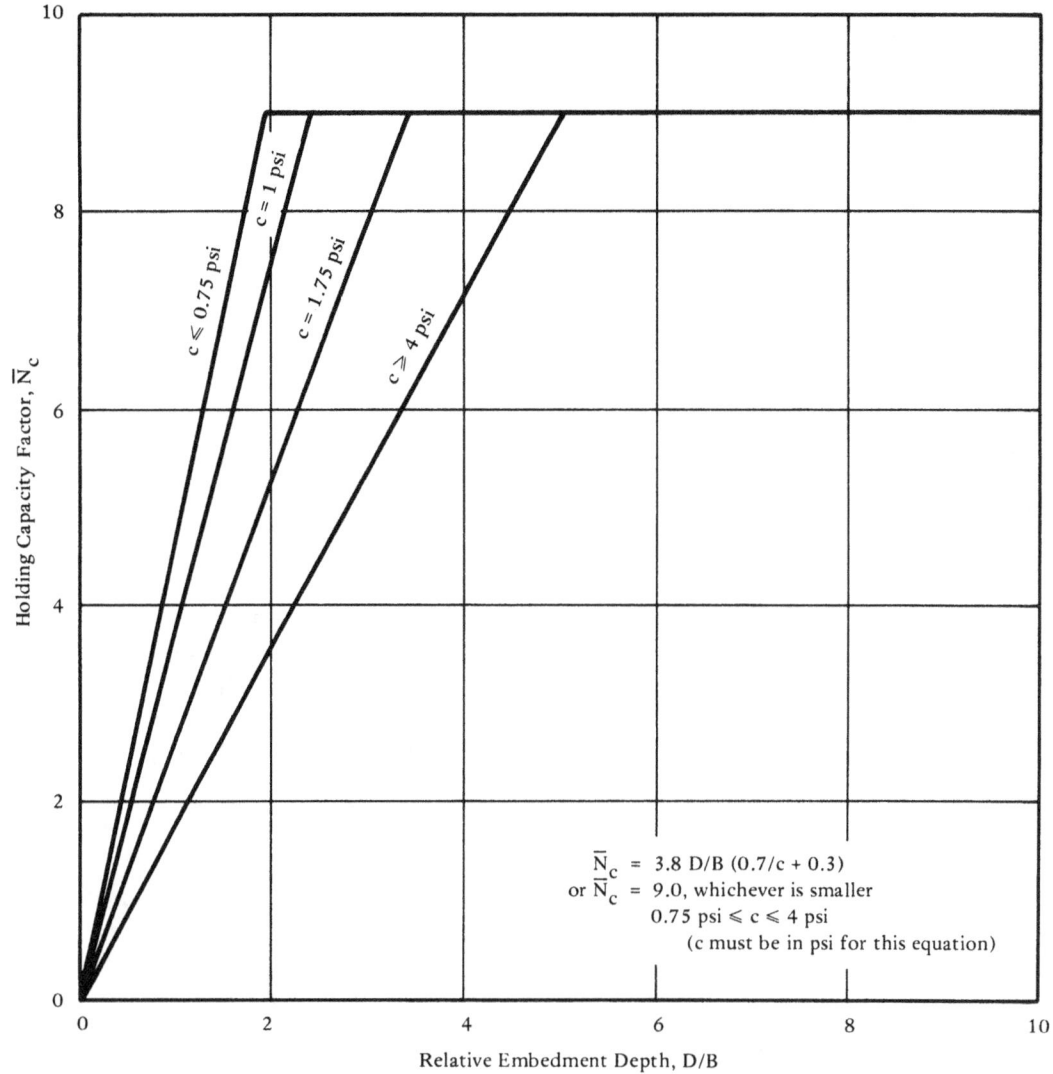

Figure 5.2-4. Design curves of holding capacity factor, $\bar{N}c$, versus relative embedment depth (D/B).

mooring system design should be developed (i.e., one which reduces effects of repeated loading) or high factors of safety (greater than 10) should be used. For other grain sizes, it is necessary to determine whether the anchor will be considered "deep" or "shallow." This can be done by referring to Figure 5.2-6 and determining whether the particular range of design parameters places **D/B** below or above the sharp breaks in the curves. If the anchor is "shallow," the design repeated-load holding capacity is one-half F_T as determined by the selected method in paragraph (4a) above. If the anchor is "deep," it is necessary to calculate the short-term holding capacity at the point where "shallow" behavior changes to "deep." The previous values of **B**, **L**, ϕ, and γ_b should be used, and paragraph (3b) should be repeated with the new **D/B**. One-half of the short-term holding capacity calculated with these parameters should be used for design purposes.

Long-term static loading. When the type of loading is long-term static, the holding capacity is F_T as determined from the method selected under paragraph (3b) above.

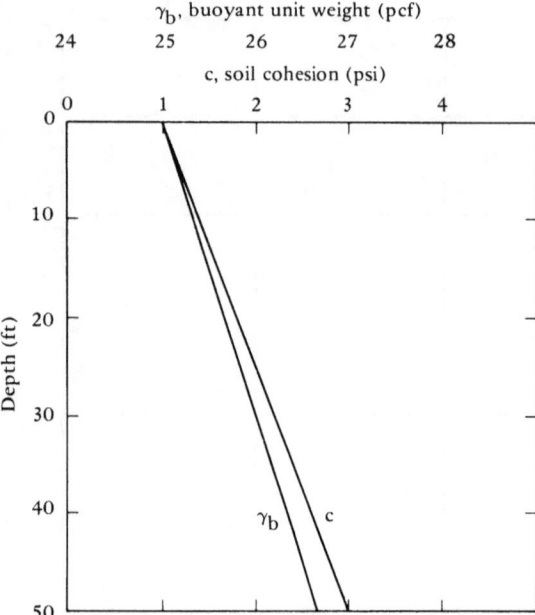

Figure 5.2-5. Recommended properties for a hypothetical cohesive soil when data on actual cohesive soil are not available.

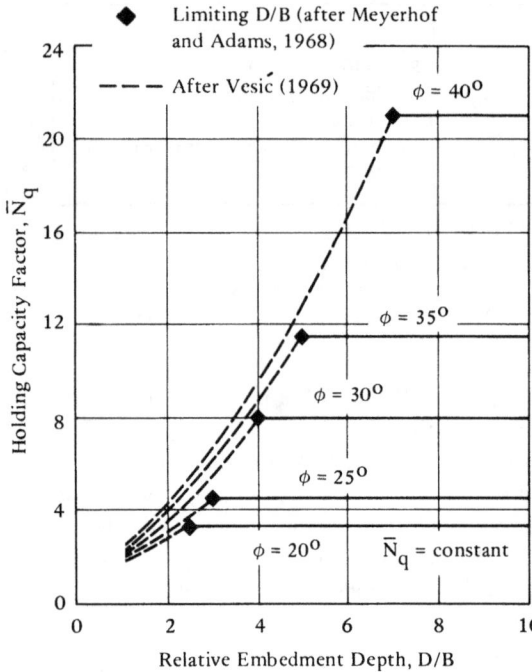

Figure 5.2-6. Holding capacity factor, $\overline{N}_q$, versus relative depth for cohesionless soil, $c = 0$.

5.3. SAMPLE PROBLEM

A Direct-Embedment Vibratory Anchor (see section 3.12) with a 3-foot-diameter fluke is to be used in a cohesive soil. The purpose of the anchor is to support a subsurface buoy that is to be in service for several years. A good quality core has been obtained, and the measured vane shear strength profile is given by the curve in Figure 5.3-1. The sensitivity of the soil is 2. The buoyant unit weight was measured and found to be about 35 pcf throughout the profile.

The penetration of the fluke must be determined first, and then the holding capacity can be estimated.

Penetration. From Figure 5.3-1 the shear strength or cohesion is shown to increase linearly with depth. Since the buoyant unit weight is constant over the soil profile, the strength can be expressed as a c/p ratio (cohesion to effective overburden pressure). At a depth of 10 feet the cohesion is equal to 2 psi or 288 psf, and the effective overburden pressure is equal to 350 psf (10 feet x 35 pcf). Therefore, the c/p ratio is equal to 0.823. The depth of penetration can be solved with Equation 5-5.

$$D = \frac{-(X+Y) \pm [(X+Y)^2 + 4W(Q + \text{Bias})]^{1/2}}{2W} \quad (5\text{-}5)$$

where $X = A_{fs}\,(c/p)\,\gamma_b$

$Y = A_{ff}\,N_c\,(c/p)\,\gamma_b$

$W = (1/2)a_s\,(1/S_t)(c/p)\,\gamma_b$

and $N_c = 9$

From Beard (1973),

$A_{fs} = 18.4 \text{ ft}^2$

$A_{ff} = 0.5 \text{ ft}^2$

$a_s = 0.813 \text{ ft}^2/\text{ft}$

$Q = 12{,}500 \text{ lb}$

Bias $= 540 \text{ lb}$

and from the soil data,

$$c/p = 0.823$$
$$\gamma_b = 35 \text{ pcf}$$
$$S_t = 2$$

Therefore,

$$X = 18.4(0.823)(35) = 530$$
$$Y = 0.5(9)(0.823)(35) = 130$$
$$W = (1/2)\,0.813\,(1/2)\,(0.823)(35) = 5.85$$

$$D = \frac{-(530 + 130) \pm [(530 + 130)^2 + 4(5.85)(12{,}500 + 540)]^{1/2}}{2(5.85)}$$

= 17.2 ft and -130 ft

Since penetrations are positive, the penetration is 17.2 feet. For this type of fluke (an eccentric keying flat plate), the ratio of keying distance to fluke length is taken as 1. The fluke keying distance is then 3 feet (1 times the fluke length). Therefore, the embedment depth to be used in the holding capacity calculations is 17.2 - 3.0 = 14.2 feet.

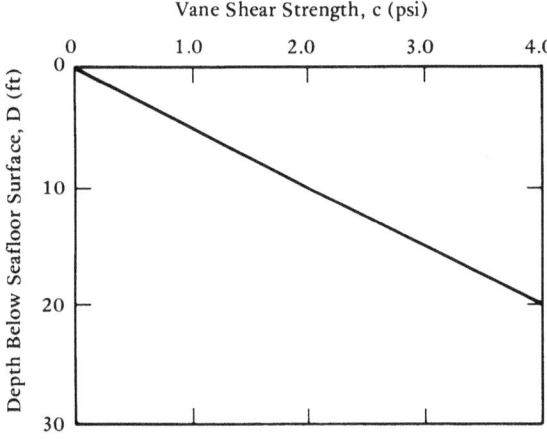

Figure 5.3-1. Vane shear strength profile for sample problem.

Holding Capacity. The step identifications by numbers in parenthesis are the same as those under section 5.2.2.

(1) Determine design parameters.

$$D = 14.2 \text{ ft}$$
$$B = 3 \text{ ft}$$
$$L = 3 \text{ ft}$$
$$A = 6.1 \text{ ft}^2 \text{ (from Beard, 1973)}$$

(2) Determine soil type.

The soil is cohesive.

(3a) Determine calculation method for cohesive soil. Core or in-situ soil data available. Therefore, the second method listed under (3a) can be used. First, the characteristic soil strength must be determined. This can be done by using Figure 5.2-3. D/B is calculated to be 14.2/3 = 4.7, and the strength at the anchor is calculated by multiplying the c/p ratio by the effective overburden pressure at that depth ($\gamma_b\,D$), which gives c_a = 0.825 x 35 x 14.2 = 409 psf. To use Figure 5.2-3, c must be in psi (409 psf x 1 psi/144 psf = 2.84 psi). From Figure 5.2-3, D_c/B is estimated to be 1.7. Multiplying by B, D_c is determined to be 5.1 feet. This is the distance above the anchor at which point the characteristic strength, c, is to be determined. At a depth of 9.1 feet (14.2 - 5.1), c is then c/p x γ_b x D = 0.823 x 35 x 9.1 = 262 psf or 1.82 psi. Now $\bar{N}_c$ can be determined from Figure 5.2-4 where D/B = 4.7 and c = 1.82; $\bar{N}_c$ = 9. $\bar{N}_q$ for cohesive soils is 1.

Now the short-term static holding capacity can be calculated from Equation 5-9.

$$F_T = A(c\,\bar{N}_c + \gamma_b\,D\,\bar{N}_q)(0.84 + 0.16\,B/L)$$
$$= 6.1\,[(262)(9) + (35)(14.2)(1)]\,[0.84 + 0.16(3/3)]$$
$$= 17{,}400 \text{ lb}$$

This is the estimated short-term static holding capacity.

(4a) Determine type of loading. The load will be applied for several years from a submerged buoy and can be considered a long-term static load.

Initially it will be a short-term static load. Therefore, the design short-term holding capacity is F_T or 17,400 pounds.

For the long-term static loading the following procedure is used. The friction angle, ϕ, was not determined by laboratory tests, and, therefore, a conservative value of 25 degrees will be used. Using $D/B = 4.7$ and $\phi = 25$ degrees, Figure 5.2-6 is used to obtain $\overline{N}_q$, which is equal to 4.5. $\overline{N}_c$ and c are equal to 0 for long-term conditions. Now Equation 5-9 can be used to find the long-term static holding capacity, F_{TD}.

$$F_{TD} = A(c\overline{N}_c + \gamma_c D\overline{N}_q)(0.84 + 0.16\,B/L)$$

$$= 6.1\,[0 + (35)(14.2)(4.5)]\,[0.84 + 0.16(3/3)]$$

$$= 13{,}600 \text{ lb}$$

F_T is larger than F_{TD}, and, therefore, F_{TD} is the design holding capacity. If the buoy were especially critical, the design holding capacity would be multiplied by 0.6 to account for possible creep effects.

Answer. The design holding capacity is 13,600 pounds.

Chapter 6. REFERENCES, BIBLIOGRAPHY, AND PATENTS

6.1. REFERENCES

Beard, R. M. (1973). Direct embedment vibratory anchor, Naval Civil Engineering Laboratory. Technical Report R-791, Port Hueneme, CA, Jun 1973. (AD 766103)

Brackett, R. L., and A. M. Parisi (1975). Development test and evaluation of a handheld hydraulic rock drill and seafloor fasteners for use by divers, Civil Engineering Laboratory, Technical Report R- , Port Hueneme, CA (to be published)

Brown, G. A., and V. A. Nacci (1971). "Performance of hydrostatic anchors in granular soils," in Preprints, Third Annual Offshore Technology Conference, Houston, TX, Apr 19-21, 1971. Dallas, TX, Offshore Technology Conference, vol 2, 1971, pp 533-542. (Paper no. OTC 1472)

Dantz, P. A., and J. B. Ciani (1967). Piston velocities of a single-impulse, deep-ocean hydrostatic ram, Naval Civil Engineering Laboratory, Technical Note N-948, Port Hueneme, CA, Dec 1967. (AD 831131)

Dantz, P. A. (1968). The padlock anchor, a fixed-point anchor system, Naval Civil Engineering Laboratory, Technical Report R-577, Port Hueneme, CA, May 1968. (AD 669113)

Frohlich, H., and J. F. McNary (1969). "A hydrodynamically actuated deep sea hard rock corer," Marine Technology Society Journal, vol 3, no. 3, May 1969, pp 53-60.

Institut Francais du Petrole, des Carburants et Lubrifiants (1970). Subseas vibro-driver, Catalog Reference 17928A, Rueil Malmaison, France, Mar 1970.

Lair, J. C. (1967). Investigation of embedding an anchor by the pulse-jet principle, Naval Civil Engineering Laboratory, Contract Report CR-68.008. Torrance, CA, Sea-Space Systems, Inc., Oct 1967. (Contract no. NBy-62225)

Meyerhof, G. G., and J. I. Adams (1968). "The ultimate uplift capacity of foundations," Canadian Geotechnical Journal, vol 5, no. 4, Nov 1968, pp 225-244.

Parisi, A. M., and R. L. Brackett (1974). Development, test and evaluation of an underwater grout dispensing system for use by divers, Civil Engineering Laboratory, Technical Note N-1347, Port Hueneme, CA, Jul 1974. (AD 786350)

Rossfelder, A. M., and M. C. Cheung (1973). Implosive anchor feasibility study, Report on Contract no. N62477-73-C-0429, Eco Systems Management Associates, La Jolla, CA, Nov 1973.

Schmid, W. E. (1969). Penetration of objects into the ocean bottom (state of the art), NCEL Contract Report CR 69.030. Princeton, NJ, W. E. Schmid, Mar 1969. (Contract no. N62399-68-C-0044). (AD 695484)

Smith, J. E. (1966). Investigation of embedment anchors for deep ocean use, Naval Civil Engineering Laboratory, Technical Note N-834, Port Hueneme, CA, Jul 1966.

Stevenson, H. J., and W. A. Venezia (1970). Jetted-in marine anchors, Naval Civil Engineering Laboratory, Technical Note N-1082, Port Hueneme, CA, Feb 1970. (AD 704488)

Taylor, R. J., and H. J. Lee (1972). Direct embedment anchor holding capacity, Naval Civil Engineering Laboratory, Technical Note N-1245, Port Hueneme, CA, Dec 1972. (AD 754745)

True, D. G. (1975). Penetration into seafloor soils, Civil Engineering Laboratory, Technical Report R-822, Port Hueneme, CA, May 1975.

Wang, M. C., V. A. Nacci, and K. R. Demars (1974). Vertical breakout behavior of the hydrostatic anchor, CEL Contract Report CR 74.005. Kingston, RI, University of Rhode Island, Feb 1974. (Contract no. N62399-72-C-0005) (AD 775658)

Vesic, Aleksandar S. (1969). "Breakout resistance of objects embedded in ocean bottom," in Civil Eng in the Oceans II, ASCE Conference Miami Beach, FL, Dec 10-12, 1969. New York, ASCE, 1970, pp 137-165.

6.2. BIBLIOGRAPHY

A. B. Chance Co. (1954). Retractable-reusable anchor, Rome Air Development Center, Griffis Air Force Base, NY, 1954. (Contract AF-30(602)-388)

A. B. Chance Co. (1969). Encyclopedia of anchoring, Bulletin 424-A, Centralia, MO, 1969.

A. C. Electronics. Defense Research Laboratories (1969). Technical Manual for Project BOMEX free-fall anchor systems, Manual no. OM69-01. Santa Barbara, CA, Feb 1969.

Adams, J. I. (1963). Uplift tests on model anchors in sand and clay, unpublished report, Research Division, the Hydro-electric Power Commission of Ontario, Canada, Sep 1963. (*Unverified*)

Adams, J. I., and D. C. Hayes (1967). "Uplift capacity of shallow foundations," Ontario Hydro Research Quarterly, vol 19, no. 1, 1967, pp 1-13.

Adams, J. I., and T. W. Klym (1972). "A study of anchorages for transmission tower foundations," Canadian Geotechnical Journal, vol 9, no. 1, Feb 1972, pp 89-104.

Aerojet-General Corporation. Proposal: Development of three types of propellant activated underwater anchors, Downey, CA, May 1961.

Aircraft Armaments, Inc. (1960). Feasibility investigation of a propellant actuated underwater anchor, Report no. ER-1966, Cockeysville, MD, Mar 1960. (Contract DA-36-034-507-ORD-3126RD) (AD 234685)

Ali, M. S. (1968). Pullout resistance of anchor plates and anchor piles in soft bentonite clay, Duke University, Soil Mechanics Series no. 17, Durham, NC, 1968, p 50. (Also MS thesis, Duke University)

American Electric Power Service Corporation, Report of anchor test, 1963. (*Unverified*)

"Anchoring and dynamic positioning," Ocean Industry, vol 1, no. 4, Aug 1966, pp 1A-16A. Special section:

> J. R. Graham. "Mooring techniques; a discussion of problems and knowledge concerning station keeping in the open sea," pp 1A-5A

> D. S. Schaner. "Wire line tension measurement; demand grows for devices to monitor loads on ships and semi-submersibles," pp 6A-10A.

> K. W. Foster. "Dynamic anchoring; man's solution to positioning vessels in deeper waters without use of anchors," pp 11A-13A.

> M. D. Korkut and E. J. Hebert. "Determining the catenary; presenting in condensed, usable form equations to find anchor chain curve," pp 14A-16A.

Baker, W. H. and R. L. Kondner (1966). "Pullout load capacity of a circular earth anchor buried in sand," Highway Research Record 108, 1966, pp 1-10.

Balla, A. (1961). "The resistance to breaking-out of mushroom foundations for pylons," in Proceedings of Fifth International Conference of Soil Mechanics Foundation Engineering, Paris, 17-22 Jul, 1961, Paris, Dunod, 1961, vol 1, pp 569-576.

Bayles, J. J., and R. E. Jochums (1965). Underwater mooring systems, Naval Civil Engineering Laboratory, Technical Note N-662, Port Hueneme, CA, Apr 1965. (AD 461146)

Bemben, S. M., and E. H. Kalajian (1969). "The vertical holding capacity of marine anchors in sand," Proceedings, Civil Engineering in the Oceans II, ASCE Conference, Miami Beach, FL, Dec 10-12, 1969. New York, American Society of Civil Engineers, 1970, pp 117-136.

Bemben, S. M., E. H. Kalajian, and M. Kupferman (1973). "The vertical holding capacity of marine anchors in sand clay subjected to static and cyclic loading," in Preprints, 1973 Offshore Technology Conference, Houston, TX, Apr 30-May 2, 1973. Dallas, TX, Offshore Technology Conference, 1973, vol 2, pp 871-880. (Paper no. OTC 1912)

Bemben, S. M., and M. Kupferman (1974). The behavior of embedded marine flukes subjected to static and cyclic loading, Report on Contract N62 399-72-C-0018. Amherst, MA, University of Massachusetts, 1974.

Bemben, S. M., M. Kupferman, and E. H. Kalajian (1971). The vertical holding capacity of marine anchors in sand and clay subjected to static and cyclic loading, Naval Civil Engineering Laboratory, Contract Report CR 72.007. Amherst, MA, University of Massachusetts, Nov 1971. (Contract no. N62399-70-C-0025) (AD 735950)

Bhatnagar, R. S. (1969). Pullout resistance of anchors in silty clay, Duke University, Soil Mechanics Series no. 18, Durham, NC, 1969, p 44. (Also MS thesis, Duke University)

Bradley, W. D. (1963). Field tests to determine the holding capacity of explosive embedment anchors, Naval Ordnance Laboratory, Technical Report 63-117, White Oak, MD, Nov 1963. (AD 422984)

Broms, B. (1965). "Design of laterally loaded piles," Proceedings American Society of Civil Engineers, Journal of the Soil Mechanics and Foundations Engineering Division, vol 91, no. SM3, May 1965, pp 79-99.

Broms, B. (1964). "Lateral resistance of piles in cohesive soils," American Society of Civil Engineers Journal Proceedings of the Soil Mechanics and Foundations Engineering Division, vol 90, no. SM2, Mar 1964, pp 27-63.

Cameron, I. (1969). "Offshore mooring devices," Petroleum Review, vol 23, no. 270, Jun 1969, pp 169-173.

Christians, J. A. (1968). Development of multi-leg mooring system, phase D. Design and layout, Army Mobility Equipment Research and Development Center, Report 1909-D, Fort Belvoir, VA, Apr 1968.

Christians, J. A., and E. P. Meisburger (1967). Development of multi-leg mooring system, phase A explosive embedment anchor, Army Mobility Equipment Research and Development Center, Report 1909-A, Fort Belvoir, VA, Dec 1967.

Christians, J. A., and E. H. Martin (1967). Development of multi-leg mooring system, phase B explosive embedment anchor fuze, Army Mobility Equipment Research and Development Center, Report 1909-B, Fort Belvoir, VA, Sep 1967. (AD 822168)

Cleveland Pneumatic Industries (1971). Embedment anchor development program, Report 4607-F, Aug 1971. (*Unverified*)

Colp, J. L., and John B. Herbich (1972). Effects of inclined and eccentric load application on the breakout resistance of objects embedded in the seafloor, Texas A&M University, Sea Grant Publication no. 72-204, May 1972. (Texas A&M University, Engineering Experimental Station, Rept no. 153-COE)

Dantz, A. (1966). Light-duty, expandable land anchor (30,000-pound class), Naval Civil Engineering Laboratory, Technical Report R-472, Port Hueneme, CA, Aug 1966. (AD 640232)

DeHart, R. C., and C. R. Ursell (1967). Force required to extract objects from deep ocean bottom, Report on Contract Nonr-336300, Southwest Research Institute, San Antonio, TX, Sep 1967, p 9.

Delco Electronics (1971). A proposal to furnish mooring systems for Project HARPOON, Proposal no. P71-44. Santa Barbara, CA, Aug 1971.

Demars, K. R., V. A. Nacci, and M. C. Wang (1972). Behavior of the hydrostatic anchors in sand, Naval Underwater Systems Center, Report on Contract no. N66604-71-C-0080, Newport, RI, May 1972. (*Unverified*)

Department of the Navy, Bureau of Yards and Docks (1962). Design manual DM-26: Harbor and coastal facilities, Washington, DC, 1962. (Superseded by 1968 ed.)

Dohner, J. A. (1966). Field tests to determine the holding powers of explosive embedment anchors in sea bottoms, Naval Ordnance Laboratory, Technical Report no. 66-205, White Oak, MD, Oct 1966.

Dowding, Charles (1970). Anchor-clay-soil interaction, Unpublished term paper, University of Illinois, Urbana, IL, 1970. (*Unverified*)

Drucker, M. A. (1934). "Embedment of poles, sheeting, and anchor piles," Civil Engineering, vol 4, no. 12, Dec 1934, pp 622-626.

Erden, S. (1971). A study of the extent of the zone of disturbance of anchors in loose soils, MS thesis, University of Massachusetts, Amherst, MA, May 1971. (*Unverified*)

Erickson, F. L. (1972). Explosive embedment anchor development program, Magnavox Company, Report No. FWD72-115, Fort Wayne, IN, Nov 1972. (Contract no. DOT-CG-04468-A)

Esquival-Diaz, R. F. (1967). Pullout resistance of deeply buried anchor in sand, Duke University (Soil Mechanics Series no. 8), Durham, NC, 1967. (Also MS thesis, Duke University)

Fotiyers, N. W., and V. A. Litkin. "Design of deep anchor plates," Soil Mechanics and Foundation Engineering, no. 5, p B-10, New York, Plenum Publishing Corp. (*Unverified*)

Fox, D. A., G. F. Parker, and V. J. R. Sutton (1970). "Pile driving into North Sea boulder clays," in Preprints, Second Annula Offshore Technology Conference, Houston, TX, Apr 22-24, 1970. Dallas, TX, Offshore Technology Conference, vol 1, 1970, pp 535-548. (Paper no. OTC 1200)

Golait, A. V. (1967). Model studies on the breaking out resistance of pile foundations with enlarge bases, Unpublished MS thesis, India Institute of Technology, Bombay, 1967. (*Unverified*)

Gordon, D. T., and R. S. Chapler (1972). Vibratory emplacement of small piles, Naval Civil Engineering Laboratory, Technical Note N-1251, Port Hueneme, CA, Dec 1972. (AD 906997)

Haley and Aldrich (Consulting Engineers) (1960). Investigations of pull-out resistance of universal ground anchors. Laconia Malleable Iron Co., Laconia, NH, File No. 60-411, 1960. (*Unverified*)

Hanna, T. H. (1968). "Factors affecting the loading behavior of inclined anchors used for the support of tie-back walls," Ground Engineering, vol 1, no. 5, Sep 1968, pp 38-41.

Harvey Alum Co. (1966). Test and evaluation of EAW-20 explosive earth anchor system, Marine Corps Landing Force Development Center, Project No. 51-52-01B, Quantico, VA, 1966. (*Unverified*)

Harvey, R. C., and E. Burley (1973). "Behavior of shallow inclined anchorage in cohesionless sand," Ground Engineering, vol 6, no. 5, Sep 1973, pp 48-55.

Healy, A. (1971). "Pullout resistance of anchors buried in sand," ASCE Proceedings, Journal of Soil Mechanics and Foundations Division, vol 97, no. SM 11, Nov 1971, pp 1615-1622.

Hollander, W. L. (1958). "Earth anchors may help you prevent pipe flotation at river crossings or in swamps," Oil and Gas Journal, vol 56, no. 21, May 26, 1958, pp 98-101.

Hollander, W. L., and R. Martin (1961). "How much can a guy anchor hold?" Electric Light and Power, vol 39, no. 5, Mar 1, 1961, pp 41-43.

Howat, M. D. (1965). The behavior of earth anchorages in sand, MS thesis, University of Bristol, 1965. (*Unverified*)

Hsieh, T. Y., and F. J. Turpin (1965). Experimental investigation of suction cup anchors, Hydronautics, Inc., Report no. TR-519-1, Sep 1965. (Contract no. Nonr-484500) (AD 803801L)

Hueckel, S. (1957). "Model tests on anchoring capacity of vertical and inclined planes," in Proceedings of the Fourth International Conference on Soil Mechanics and Foundation Engineering, London, 12-24 Aug 1957. London, Butterworths Scientific Publications, vol 2, 1957, pp 203-206.

Johnson, V. E., R. J. Etter, and F. J. Turpin (1967). "Suction cup anchors for underwater mooring and handling," paper presented at American Society of Mechanical Engineers, Petroleum Mechanical Engineering Conference, Philadelphia, PA, Sep 17-20, 1967.

Kalajian, E. H. (1971). The vertical holding capacity of marine anchors in sand subjected to static and cyclic loading, Ph D thesis, University of Massachusetts, Amherst, MA, 1971.

Kalajian, E. H., S. M. Bemben (1969). The vertical pullout capacity of marine anchors in sand, Report no. UM-69-5, University of Massachusetts, School of Engineering, Amherst, MA, Apr 1969. (AD 689522)

Kanayan, A. S. (1963). "Analysis of horizontally loaded pipes," Soil Mechanics and Foundation Engineering, no. 2, Mar-Apr 1963. (*Unverified*)

Karafiath, L., and M. G. Bekker (1957). An investigation of gun anchoring spades under the action of impact loads, Army Tank-Automotive Command, Report no. 19, Warren, MI, Oct 1957. (AD 156419)

Kennedy, J. L. (1969). "This lightweight, explosive-set anchor can stand a big pull," Oil and Gas Journal, vol 67, no. 16, Apr 21, 1969, pp 84-86.

Khadilkar, B. S., A. K. Paradkar, and Y. S. Golait (1971). "Study of rupture surface and ultimate resistance of anchor foundations," in Proceedings of the Fourth Asian Regional Conference on Soil Mechanics and Foundation Engineering, Bangkok, 26 Jul-1 Aug 1971. Bangkok, Asian Institute of Technology, 1971, vol 1, pp 121-127; discussion, vol 2, pp 139-140.

Kupferman, M. (1971). The vertical holding capacity of marine anchors in clay subjected to static and cyclic loading, MS thesis, University of Massachusetts, Amherst, MA, 1971. (*Unverified*)

Kwasniewski, J., and L. Sulikowska (1964). "Model investigations on anchoring capacity of vertical cylindrical plates," in Proceedings of the Seminar on Soil Mechanics and Foundation Engineering, Lodz, 1964. (*Unverified*)

Langley, W. S. (1967). Uplift resistance of groups of bulbous piles in clay," MS thesis, Nova Scotia Technical College, Halifax, NS, 1967. (*Unverified*)

Larnach, W. J. (1972). "Pullout resistance on inclined anchors," Ground Engineering, vol 5, no. 4, Jul 1972, pp 14-17.

Lee, H. J. (1972). Unaided breakout of partially embedded objects from cohesive seafloor soils, Naval Civil Engineering Laboratory, Technical Report R-755, Port Hueneme, CA, Feb 1972. (AD 740751)

Liu, C. L. (1969). Ocean sediment holding strength against breakout of embedded objects, Naval Civil Engineering Laboratory, Technical Report R-635, Port Hueneme, CA, Aug 1969. (AD 692411)

MacDonald, H. F. (1963). Uplift resistance of caisson piles in sand, MS thesis, Nova Scotia Technical College, Halifax, NS, 1963. (*Unverified*)

Magnavox Co. Explosive embedment penetrometer system, Final Report No. TP 4912. Fort Wayne, IN. (*Unverified*)

Magnavox Co., Government and Industrial Division, Self embedment anchor developments. Urbana, IL. (*Unverified*)

Mardesich, J. A., and L. R. Harmonson (1969). Vibratory embedment anchor system, Naval Civil Engineering Laboratory, Contract Report CR-69.009. Long Beach, CA, Ocean Science and Engineering, Inc., Feb 1969. (Contract no. N62399-68-C-0008) (AD 848920L)

Mariupolskii, L. G. (1965). The bearing capacity of anchor foundations, Soil Mechanics and Foundation Engineering, vol 3, no. 1, Jan-Feb 1965, pp 26-32. (*Unverified*)

Markowsky, M., and J. I. Adams (1961). "Transmission towers anchored in muskeg," Electrical World, vol 155, no. 8, Feb 20, 1961, pp 36-37, 68.

Matlock, M., and L. C. Reese (1962). Generalized solutions for laterally loaded piles, ASCE Transactions, vol 127, pt I, 1962, pp 1220-1251.

Matsuo, M. (1967). "Study on the uplift resistance of footing (I)," Soils and Foundations, Japan, vol 7, no. 4, Dec 1967, pp 1-37.

Matsuo, M. (1968). "Study on the uplift resistance of footing (II)," Soils and Foundations, Japan, vol 8, no. 1, Mar 1968, pp 18-48.

Mayo, H. C. (1972). "Rapid mooring-construction system," The Military Engineer, vol 64, no. 418, Mar-Apr 1972, pp 110-111.

Mayo, H. C. (1973a). Explosive embedment anchors for ship mooring, Army Mobility Equipment Research and Development Center, Report 2078, Fort Belvoir, VA, Nov 1973.

Mayo, H. C. (1973b). "Explosive anchors for ship mooring," Marine Technology Society Journal, vol 7, no. 6, Sep 1973, pp 27-34.

McKenzie, R. J. (1971). "Uplift testing of prototype transmission tower footings," in Proceedings of First Australian-New Zealand Conference on Geomechanics, Melbourne, Aug 1971, vol 1, pp 283-290.

Meyerhof, G. G. (1973). "The uplift capacity of foundations under oblique loads," Canadian Geotechnical Journal, vol 10, no. 1, Feb 1973, pp 64-70.

Migliore, H. J., and H. J. Lee (1971). Seafloor penetration tests: Presentation and analysis, Naval Civil Engineering Laboratory, Technical Note N-1178, Port Hueneme, CA, Aug 1971. (AD 732369)

Muga, B. J. (1967). "Bottom breakout forces," in Proceedings Civil Engineering in the Oceans, ASCE Conference, San Francisco, CA, Sep 6-8, 1967, pp 569-600. New York, American Society of Civil Engineers, 1968, pp 569-600.

Muga, B. J. (1966). Breakout forces, Naval Civil Engineering Laboratory, Technical Note-863, Port Hueneme, CA, Sep 1966, p 24.

Muga, B. J. (1968). Ocean bottom breakout forces, including field test data and the development of an analytical method, Naval Civil Engineering Laboratory, Technical Report R-591, Port Hueneme, CA, Jun 1968, p 140 (AD 837647)

Neely, W. J. (1972). "Sheet pile anchors: Design reviewed the importance of flexibility in the design of sheet pile anchors in sand," Ground Engineering, vol 5, no. 3, May 1972, pp 14-16.

North American Aviation, Inc. (1965). Hydrostatic embedment anchor tests, Summary Report NA65H-387, Columbus, OH, 1965. (*Unverified*)

Offshore/Sea Development Corp. (1969). Technical proposal: Mud and rock anchor feasibility and design study, 1969.

Radhakrishna, A. S., and S. I. Adams (1973). "Long-term uplift capacity of augered footings in fissured clay," Canadian Geotechnical Journal, vol 10, no. 4, Nov 1973, pp 647-652.

Raecke, D. A., and H. J. Migliore (1971). Seafloor pile foundations: State-of-the-art and deep-ocean emplacement concepts, Naval Civil Engineering Laboratory, Technical Note N-1182, Oct 1971. (AD 889087)

Redick, T. E. (1962). Anchor study, phase 1A, Naval Air Engineering Laboratory, Naval Air Materiel Center, Report NAEL-ENG-6853, Philadelphia, PA, Apr 1962.

Ridgeway, J. J. (1970). "Explosive anchors for sea mooring," Undersea Technology, vol 11, no. 12, Dec 1970, pp 16-17.

Robinson, F. S., and J. A. Christians (1967). Development of multi-leg mooring system, phase C. Installation, Army Mobility Equipment Research and Development Center. Report 1909-C, Fort Belvoir, VA, Oct 1967.

Schmidt, B., and J. P. Kirtstensen (1964). "The pulling resistance of inclined anchor piles in sand," Danish Geotechnical Institute Bulletin no. 18, 1964.

Schuette, H. W., and P. E. Sweeney (1969). Mud and rock anchor feasibility and design study for multi-leg mooring system; final report on Phase 1, AAI Corp. Report no. ER-5885, Cockeysville, MD, Sep 1969. (Contract no. DAAK02-69-C-0554) (AD 860147L)

Sherwood, W. G. (1967). "Developing a free-fall, deep-sea mooring system," in Transactions of the Second International Buoy Technology Symposium, Washington, DC, Sep 18-20, 1967. Washington, DC, Marine Technology Society, 1967, pp 19-35.

Smith, J. E., and P. A. Dantz (1963). A perspective on anchorages for deep ocean constructions, Naval Civil Engineering Laboratory, Technical Note N-552, Port Hueneme, CA, Dec 1963. (AD 426202)

Smith, J. E. (1971). Explosive anchor for salvage operations; progress and status, Naval Civil Engineering Laboratory, Technical Note N-1186, Port Hueneme, CA, Oct 1971. (AD 735104) Smith, J. E. (1972). Explosive anchor for salvage operations; progress and status, addendum, TN-1186A, Jan 1972.

Smith, J. E. (1965). Structures in deep ocean engineering manual for underwater construction, chap 7. Buoys and anchoring systems, Naval Civil Engineering Laboratory, Technical Report R-284-7, Port Hueneme, CA, Oct 1965. (AD 473928)

Smith, J. E. (1955). Evaluation of the EZY Pier-Anchor, Naval Civil Engineering Laboratory, Technical Note N-204, Port Hueneme, CA, Feb 1955. (AD 81220L)

Smith, J. E. (1966a). "Investigation of embedment anchors for deep ocean use," paper presented at American Society of Mechanical Engineers, 66-PET-32.

Smith, J. E. (1966b). Investigation of free-fall embedment anchor for deep ocean application, Naval Civil Engineering Laboratory, Technical Note N-805, Port Hueneme, CA, Mar 1966. (AD 808818L)

Smith, J. E. (1954). Stake pile development for moorings in sand bottoms, Naval Civil Engineering Laboratory, Technical Note N-205, Port Hueneme, CA, NOV 1954. (AD 81261)

Smith, J. E. (1963). Umbrella pile-anchors, Naval Civil Engineering Laboratory, Technical Report R-247, Port Hueneme, CA, May 1963, (AD 408404)

Smith, J. E., R. M. Beard, and R. J. Taylor (1970). Specialized anchors for the deep sea; progress summary, Naval Civil Engineering Laboratory, Technical Note N-1133, Port Hueneme, CA, Nov 1970. (AD 716408)

Smith, J. E. (1957). Stake pile tests in mud bottom, Naval Civil Engineering Laboratory, Letter Report L-022, Port Hueneme, CA, Sep 1957.

Sowa, V. A. (1970). "Pulling capacity of concrete cast in situ bored piles," Canadian Geotechnical Journal, vol 7, no. 4, Nov 1970, pp 482-493.

Spence, W. M. Uplift resistance of piles with enlarged base in clay, MS thesis, Nova Scotia Technical College, Halifax, NS. (*Unverified*)

Stevenson, H. J., and W. A. Venezia (1970). Jetted-in marine anchors, Naval Civil Engineering Laboratory, Technical Note N-1082, Port Hueneme, CA, Feb 1970. (AD 704488)

Sutherland, H. B. (1965). "Model studies for shaft raising through cohesionless soils," in Proceedings of the Sixth International Conference on Soil Mechanics and Foundation Engineering, Montreal, 8-15 Sep 1965. Toronto, University of Toronto Press, 1965, vol 2, pp 410-413.

Taylor, R. J., and R. M. Beard (1973). Propellant-actuated deep water anchor; interim report, Naval Civil Engineering Laboratory, Technical Note N-1282, Port Hueneme, CA, Aug 1973. (AD 765570)

Techniques Louis Menard. Publication P/95: Mooring anchors, Longjumeau, France, 1970. (*Unverified*)

Thomason, R. A. (1964). Propellant-actuated embedment anchor system, Report on Contract no. P.O. 127/34, Downey, CA, Aerojet-General Corp., Jun 1964.

Thomason, R. A. (1968). Propellant-actuated embedment anchor, Naval Civil Engineering Laboratory, Contract Report CR 69.026. Downey, CA, Aerojet-General Corp., Nov 1968. (Report no. AGC-3324-01(01)FP) (Contract no. N62399-68-C-0002) (AD 850896)

Thomason, R. A., and H. W. Wedaa (1961). Special report: The Chuckawalla anchor, Aerojet-General Corp., Report no. 1327-61(02)PB, Downey, CA, Jun 1961.

Timar, J. G., and S. M. Bemben (1973). Thy influence of geometry and size on the static vertical pull-out capacity of marine anchors embedded in very loose, saturated sand, University of Massachusetts, School of Engineering. Report no. UN 73-3, Amherst, MA, Mar 1973. (AD 761623)

Tolson, B. E. (1970). A study of the vertical withdrawal resistance of projectile anchors, MS thesis, Texas A&M University, College Station, TX, May 1970.

Trofimenkov, J. G., and L. G. Mariupolskii (1965). "Screw piles for mast and tower foundations," in Proceedings of the Sixth International Conference on Soil Mechanics and Foundation Engineering, Montreal, 8-15 Sep 1965. Toronto, University of Toronto Press, 1965, vol 2, pp 328-332.

True, D. G. (1974). "Rapid Penetration into Seafloor Soils," in Preprints, Offshore Technology Conference, Houston, TX, May 6-8, 1974. Dallas, TX, Offshore Technology Conference, 1974, vol 2, pp 607-618. (Paper no. OTC 2095)

True, D. G., J. A. Drelicharz, and J. E. Smith. Deep water anchor expedient mooring system, Civil Engineering Laboratory, Technical Note N-   , Port Hueneme, CA. (To be published)

Turner, E. A. (1962). "Uplift resistance of transmission tower footings," ASCE Proceedings, Journal of the Power Division, vol 88, no. P02, Jul 1962, pp 17-33.

Wilson, S. D., and D. E. Hilts (1967). "How to determine lateral load capacity of piles," Wood Preserving, Jul 1967. (*Unverified*)

Wiseman, R. J. (1966). Uplift resistance of groups of bulbous piles in sand, MS thesis, Nova Scotia Technical College, Halifax, NS, 1966. (*Unverified*)

Yilmaz, M. (1971). The behavior of groups of anchors in sand, PhD thesis, University of Sheffield, England, 1971. (*Unverified*)

6.3. PATENTS

Anderson, M. H., Explosive operated anchor assembly, U.S. Patent No. 3,207,115, filed June 17, 1963.

Anderson, V. C., et al., Pressure actuated anchor, U.S. Patent No. 3,311,080.

Bayles, J. J., Apparatus for mooring instruments at a predetermined depth, U.S. Patent No. 3,471,877.

Bauer, R. F., et al., Anchoring method and apparatus, U.S. Patent No. 2,891,770.

Costello, R. B., et al., Dynamic anchor, U.S. Patent No. 3,187,705.

Edwards, T. B., Vibratory sea anchor driver, U.S. Patent No. 3,417,724, filed Sep 27, 1967.

Ewing, W. M., et al., Deep-sea anchor, U.S. Patent No. 2,703,544, Mar 8, 1955.

Feiler, A. M., Embedment anchor, U.S. Patent No. 3,032,000, May 1, 1962.

Halberg, P. V., et al., Mooring apparatus, U.S. Patent No. 3,291,092, Dec 13, 1966.

Mott, G. E., et al., Deep water anchor, U.S. Patent No. 3,411,473.

Mott, G. E., et al., Method for installing a deep water anchor, U.S. Patent No. 3,496,900, Feb 24, 1970.

Pannell, O. R., Explosive embedment rock anchor, U.S. Patent No. 3,431,880, Mar 11, 1969.

Parson, E. W., et al., Explosive center hole anchor, U.S. Patent No. 3,401,461, Sep 1966.

Thomason, R. A., et al., Embedment anchor, U.S. Patent No. 3,154,042.

Vincent, R. P., Explosively driven submarine anchor, U.S. Patent No. 3,525,187, Aug 25, 1970.

Appendix A

SUPPLEMENTARY TABULAR DATA ON SPECIFIC ANCHOR DESIGNS

Table A-1 provides a summary of characteristics of uplift-resisting anchors, including such items as operational and performance characteristics and cost. Holding capacity and penetration data are presented in Tables A-2 and A-3, respectively. The number of data points is inconsistent between Tables A-2 and A-3 because in some tests penetration depth was not measured while in others holding capacity was not or could not be recorded.

Type of Anchor	Agency	Anchor	Approximate Size (ft)	Approximate Weight (lb)		Method of Positioning
				System	Anchor Only	
Propellant-Actuated Anchors	Magnavox Company	Embedment Anchor System (Model 1000)	3-1/2 (ht)	[25]	3	free-fall
		Model 2000	4 (ht)	[65]	6.75	controlled lowering
	Edo Western Corp.	VERTOHOLD Embedment Anchor (Model 10K)	2-1/2 (ht)	60	25	controlled lowering
	Teledyne Movible Offshore, Inc.	SEASTAPLE Explosive Embedment Anchor, Mark 5	2 to 3 (ht)	60	10	controlled lowering
		Mark 50	8 to 10 (ht)	1,900	250	controlled lowering
	U.S. Navy (CEL)	CEL 20K Propellant Anchor	12 (ht)	1,500	300 to 500	controlled lowering
		Navy 100K Propellant Anchor	12 (ht)	15,000	1,300 to 2,500	controlled lowering
	U.S. Army (MERDC)	Explosive Embedment Anchor, Model XM-50	9 (ht)	1,850	225	controlled lowering
		Model XM-200	13 (ht)	5,300	900	controlled lowering

Table A-1. Summary of Characteristics of Uplift Resisting Anchors

Method of Activation	Intended Operational Depths			Advertised Nominal Rated Holding Capacity[a] (kips)	Estimated Short-Term Holding Capacity[b] (kips)	
	Minimum (ft)	Maximum (ft)	Maximum Experienced (ft)		Soft Clay (mud)	Sand
bottom contact firing	10	10,000	13,700	1	0.5	2.0
bottom contact firing or remote manual firing	10	—	42	2	0.8	2.0
bottom contact firing or remote manual firing	30	diver depth limit	1,100	10	[2]	10
bottom contact firing or remote control firing	10	1,000	6,000	5	0.5 to 2.5	2 to 10
bottom contact firing or remote control firing	50	1,000	—	50	17 to 26	40
bottom contact firing	50	20,000	18,600	20	19 to 40	[40 to 60]
remote control firing	50	500	700	100	[50 to 100]	[150 to 250]
bottom contact firing or remote control firing	25	150	45	50	15 to [30]	30 to 75
bottom contact firing or remote control firing	40	150	55	200	30 to 85	77 to 280

Remarks
1. Maximum allowable load on flukes = 2,000 lb. 2. Anchors are fabricated on order; size adjustment is possible. 3. Two nose configurations - one for rock, one for sediments. 4. Principal objective - readily placed, light-duty mooring system; any depth. 5. USCG presently refining anchor for shallow water.
1. The Navy noted difficulty with fluke-keying in clay. 2. Anchors are fabricated to order. 3. Propellant load varies with different seafloors. 4. Anchor has been reportedly used.
1. The Navy noted structural weakness in flukes that can be corrected. 2. Company primarily deals in services; will furnish and place anchors on order. 3. Placement of more than 100 anchors has been reported.
1. Anchor is not yet a production item, but it can be used by the Government and Industry. 2. Design objective is for a low-cost system that uses expendable components for deep-water applications. 3. Two anchor configurations, one for rock and one for sediments; three sizes of sediment fluke are used. 4. Twelve anchors have been emplaced; tests continuing.
1. Anchor is not a production item, but it can be used by the Government and Industry. 2. Design objective is for a high-capacity anchor for salvage situations where conventional anchors do not function reliably, e.g., on coral bottoms. 3. Two fluke configurations are used: one for coral and rock, and one for sediments.
1. Design objective is for an easily installable, high-capacity anchor for shallow-water mooring of large army tanker supply ships. 2. Anchors have been extensively tested and have been used successfully in actual ship moorings.

continued

Type of Anchor	Agency	Anchor	Approximate Size (ft)	Approximate Weight (lb)		Method of Positioning
				System	Anchor Only	
Propellant-Actuated Anchors (cont)	Union Industrielle at d'Enterprise (UIE)	PACAN Model 3DT	25 (ht)	5,300	—	controlled lowering
		Model 10DT	44 (ht)	19,400	—	controlled lowering
Vibratory Anchors	U.S. Navy (CEL)	Direct-Embedment Vibratory Anchor	17 to 19 (ht)	1,700 to 2,000	100 to 400	controlled lowering
	Ocean Science and Engineering, Inc.	Vibratory Embedment Anchor, Model 1	—	150	—	controlled lowering
		Model 2000	46 (ht)	1,000	200	controlled lowering
Screw-In Anchors	Anchoring, Inc.	Chance Special Offshore Multi-Helix Screw Anchor, Model 3-6"	15 (ht) (shaft can be lengthened to 100 ft by extra signs)	6,000	70	controlled lowering
		Model 2-6"	—	6,000	65	controlled lowering
		Model 3-4"	—	6,000	60	controlled lowering
Driven Anchors	U.S. Navy (NAVFAC)	Stake Pile, Class C 8-in.	30 (lth)	[2,400 +]	1,400	controlled lowering, pile follower
		Class B 12-in.	30 (lth)	[3,600 +]	2,600	controlled lowering, pile follower
		Class A 16-in.	30 (lth)	[4,500 +]	1,400	controlled lowering, pile follower

132A

Table A-1. Continued

Method of Positioning	Method of Activation	Intended Operational Depths			Advertised Nominal Rated Holding Capacity[a] (kips)	Estimated Short-Term Holding Capacity[b] (kips)	
		Minimum (ft)	Maximum (ft)	Maximum Experienced (ft)		Soft Clay (mud)	Sand
controlled lowering	bottom contact firing	—	20,000	300	66	[20 to 30]	[30 to 70]
controlled lowering	bottom contact firing	—	3,000	—	220	[40 to 80]	[100 to 200]
controlled lowering	battery, electric	[1]	6,000	2,500	40	[15 to 30]	[40 to 60]
controlled lowering	—	[1]	—	5	10	[3]	[10]
controlled lowering	diesel hydraulic motor	[1]	500	50	80	27	80 to 120
controlled lowering	diesel hydraulic motor/diver-operated impact wrench	0	1,000	325	—	8-12	—
controlled lowering	diesel hydraulic motor/diver-operated impact wrench	0	1,000	—	—	[5]	—
controlled lowering	diesel hydraulic motor/diver-operated impact wrench	—	—	—	—	[5]	—
controlled lowering, pile follower	drop hammer	0	300	—	100	[10/50]	[50/150]
controlled lowering, pile follower	drop hammer	0	300	—	200	[15/70]	[50/300]
controlled lowering, pile follower	drop hammer	0	300	—	300	[20/90]	[50/300]

Hardware Cost Per Anchor Installation[c] ($)		Remarks
Installation Mechanism Recovered[d]	Installation Mechanism Expended[d]	
4,570[e]	impractical	1. Thirty anchors have been made with the 3DT. Model 10DT has not yet been tested. 2. Anchors are not stocked, but they can be fabricated on order. 3. Three plate-type anchors and one spike-type anchor have been designed for adaptation to different seafloors.
12,570[e]	impractical	
4,000 (approx)	10,000 (approx)	1. Anchor not in production, but it can be used by the Government and Industry. 2. Present anchor is second generation design. 3. Three fluke sizes are available for different types of seafloors. 4. Design objectives are for a safe lightweight, low-cost consistently high-capacity uplift-resisting anchor.
–		1. Model 1 is under test; Model 2000 is a second generation stock item. 2. Design objectives for the Model 2000 are for a safe, lightweight, low-cost, consistently high-capacity uplift-resisting anchor.
3,184[e]		
375/pair installed (approx)	–	1. Torsional strength of shaft may limit capacity in high-strength soils. 2. Anchors are stock items. 3. Larger helix diameters up to 15 inches are available. 4. Anchors have been used extensively in pairs as pipeline anchors. 5. Size explanation: 3-6" means three helixes, 6 in. in diam.; helix spacing is 2 to 3 ft. 6. Only three of a multitude of sizes and types were chosen; anchors with capacities to 40 kips are available.
2,500	–	1. All models are open-ended steel pipes with fins extending along upper 40% of length. 2. Design objective is for a fixed-point multidirectional anchor for Fleet moorings. 3. System weight depends on length of follower required. 4. In estimating holding capacity column, first value is for uplift and second is for load applied at about a 4 to 5-degree angle to top of pile driven about 5 ft below seafloor.
3,100	–	
3,600	–	

Type of Anchor	Agency	Anchor	Approximate Size (ft)	Approximate Weight (lb)		Method of Positioning	Method of Activation
				System	Anchor Only		
Driven Anchors (cont)	U.S. Navy (NAVFAC)	Umbrella Pile-Anchor, Mark III	9 (ht)	[2,400 +]	1,400	controlled lowering, pile follower	drop hammer
		Mark IV	8 (ht)	[3,600 +]	2,200	controlled lowering, pile follower	drop hammer
Deadweight-Type Anchors	Delco Electronics	Free-Fall Anchor System, smallest size	[4.5]	600	600	free-fall	N/A
		Largest size	[13]	24,000	24,000	free-fall	N/A

NOTE: Data in brackets is estimated and based on the best judgment of the authors. It is presented where possible to provide some reasonable guide to size, capacity, or shape. Due to lack of data, the techniques Louis Menard Rotating Plate Anchor and Expanded Rock Anchor are not included in this table. See Sections 3.18 and 3.19.

[a] See Table A-2 for additional data and Appendix B for calculated capacities.

[b] See Table A-2 for the limited available data on holding capacities in rock.

[c] See Chapter 3 for additional cost information.

[d] Installation mechanism refers to, e.g., gun assembly, drive assembly, pile driver, etc.

[e] Costs based upon ten installations, where installation mechanism is amortized over varying numbers of installations, depending upon mechanism type.

Table A-1. Continued

Intended Operational Depths			Advertised Nominal Rated Holding Capacity[a] (kips)	Estimated Short-Term Holding Capacity[b] (kips)		Hardware Cost Per Anchor Installation[c] ($)	
Minimum (ft)	Maximum (ft)	Maximum Experienced (ft)		Soft Clay (mud)	Sand	Installation Mechanism Recovered[d]	Installation Mechanism Expended[d]
0	300	50	300	[50 to 100]	[250]	4,500	
0	300	50	300	[50 to 100]	[250]	7,500	
[500]	[10,000]	—	0.5	0.5	0.5	—	600
[500]	[20,000]	[18,000]	22	22	22	—	30,000

	Remarks
	1. Design objective is for a direct-embedment, high-capacity bearing- and uplift-resisting pile anchor. 2. System weight depends on length of follower required. 3. Some structural failures of the anchor justify a reduction in rated capacity to 250 kips.
	1. Many intermediate sizes are available by virtue of modular construction. 2. Holding capacities will differ for loadings other than direct uplift.

133C

Agency	Anchor	Soft Clay (Mud)		
		No. of Tests	Maximum Vertical Load (kips)	
			Pull-Out[a]	No Pull-Out[b]
Magnavox	Model 1000[c]	—	—	0.5
	Model 2000[c]	—	—	0.8
Edo Western	VERTOHOLD 10K	1	6	—
		1	—	—
Teledyne Movible Offshore, Inc.	SEASTAPLE, Mark 5	5	0.5 to 2.5	—
	SEASTAPLE, Mark 50	2	17 to 26	—
U.S. Navy (CEL)	20K Propellant Anchor	3	8 to 20[d]	—
		5	19 to 40	—
	100K Propellant Anchor	4	58 to 92	—
		1	—	92[f]
U.S. Army (MERDC)	Model XM-50	1	15	—
	Model XM-200	10	30 to 85	—
Union Industrielle et d'Enterprise	PACAN 3DT	—	—	—
U.S. Navy (CEL)	Vibratory Anchor	2	5 to 52	—
Ocean Science and Engineering, Inc.	Model 1	—	—	—
	Model 2000	—	27	—
Anchoring, Inc.[h]	Model 2-6"	1	5.5	—
	Model 3-4"	—	—	—
	Model 3-6"	3	10.5 to 12	—

Table A-2. Holding Capacity Data

Medium to Stiff Clay			Sand or Sand and Gravel			Coral and Rock		
No. of Tests	Maximum Vertical Load (kips)		No. of Tests	Maximum Vertical Load (kips)		No. of Tests	Maximum Vertical Load (kips)	
	Pull-Out[a]	No Pull-Out[b]		Pull-Out[a]	No Pull-Out[b]		Pull-Out[a]	No Pull-Out[b]
–	–	0.8	–	–	2	–	–	2
–	–	3.5	–	–	2.5	–	–	1.5
–	–	–	2	10.7 to 11	–	1 (rock)	–	16
–	–	–	–	–	–	–	–	–
11	3.2 to 9	–	11	1.4 to 10	–	6 (coral)	2.3 to 6	–
1	–	7.5	2	–	7.5 to 15	–	–	–
–	–	–	1	41	–	1 (shale)	74	–
–	–	–	3	–	25[e]	–	–	–
1	40	–	4	27 to 48	–	4 (rock)	20 to 107	–
–	–	–	1	–	55	–	–	–
–	–	–	1	65[f]	–	3 (coral)	65 to 120	–
–	–	–	1	–	130	4 (coral)	–	75 to 150
1	–	230	–	–	–	2 (rock)	45 to 64	–
–	–	–	–	–	–	1 (rock)	–	168
2	50	–	4	–	56 to 75	2 (coral)	65 to 80	–
4	–	45 to 60	5	30 to 72	–	–	–	–
15	36 to 250	–	3	140 to 220	77 to 282	7 (coral)	60 to 220	–
4	70 to 130	77 to 130	4	–	–	–	–	–
–	–	–	–	–	–	30 (coral)	110,000	–
6	9 to 62	–	6	14 to 70	–	–	–	–
–	–	–	1	–	62	–	–	–
2	5.6 to 9.6	–	–	–	–	–	–	–
1	–	17	–	–	–	–	–	–
–	75 to 96	–	–	88 to 134	–	–	–	–
–	–	–	–	–	–	–	–	–
1	9.5	–	–	–	–	–	–	–
1	12	–	–	–	–	–	–	–

continued

Table A-2. Continued

Agency	Anchor	Soft Clay (Mud)			Medium to Stiff Clay		
		No. of Tests	Maximum Vertical Load (kips)		No. of Tests	Maximum Vertical Load (kips)	
			Pull-Out[a]	No Pull-Out[b]		Pull-Out[a]	No Pull-Out[b]
U.S. Navy (NAVFAC)	Stake Pile						
	Class C 8-in.	2	95	---	---	---	---
	Class B 12-in.	2	73	---	---	---	---
	Class A 16-in.	2	45	---	---	---	---
	Umbrella Pile						
	Mark III	---	---	---	---	---	---
	Mark IV	2	135 to 152	---	---	---	---
		---	---	---	---	---	---

[a] Anchor was pulled out intact.

[b] Anchor was not pulled out either from anchor or riser failure; or anchor was proof-tested and left in service.

[c] Magnavox Co. has done extensive testing in simulated laboratory conditions and in on-site situations. The exact number of tests is not known. The figures listed are approximations based on graphs and other data provided by the company.

[d] Small anchor fluke (sand fluke) used; flukes have one-third the area of the normal clay fluke.

[e] Anchors proof-tested to this load; anchors are being used in an installation.

[f] New plate-like flukes not used; original umbrella flukes used.

[g] Tests run in mud with a sand cover in Chesapeake Bay.

[h] Considerable data available concerning performance of anchors in terrestrial soils.

Sand or Sand and Gravel		No. of Tests	Coral and Rock	
Maximum Vertical Load (kips)			Maximum Vertical Load (kips)	
Pull-Out[a]	No Pull-Out[b]		Pull-Out[a]	No Pull-Out[b]
—	—	—	—	—
300 to 355	.....	.....	....	....
160	—	—	—	—
350	—	—	—	—
353	—	—	—	—
.....	300	.....	....	—

Table A-3. Test Penetration Data

Agency	Anchor	Soft Clay (Mud)		Medium to Stiff Clay		Sand or Sand/Gravel		Coral and Rock	
		No. of Tests	Penetration (ft)	No. of Tests	Penetration (ft)	No. of Tests	Penetration (ft)	No. of Tests	Penetration (ft)
Magnavox	Model 1000[a]	–	16	–	10 (est)	–	9	–	1.3
	Model 2000[a]	–	16	–	10 (est)	–	9	–	1.3
Edo Western	VERTOHOLD 10K	1	10	–	–	14	9 to 17	21 (coral)	3 to 7
Teledyne Movible Offshore, Inc.	SEASTAPLE Mark 5	2	24 to 25	8	9 to 14	9	2 to 13	6 (coral)	1 to 5
	SEASTAPLE Mark 50	2	40 to 45	–	–	1	30	1 (rock)	5
U.S. Navy (CEL)	20K Propellant Anchor	4	34 to 47	–	–	5	9-1/2[b] to 30	4 (rock)	2 to 3
	100K Propellant Anchor	5[c]	34 to 54	–	–	3[c]	5 to 18	3 (rock)	4 to 5
U.S. Army (MERDC)	Model XM-50	1	41	1	40	5	16 to 26	2 (coral)	18 to 23
	Model XM-200	7	14 to 42	19	16 to 49	8	19 to 30	5 (coral)	10 to 21
U.S. Navy (CEL)	Vibratory Anchor	2	7 to 15	7	4 to 16	8	2 to 16	–	–
Ocean Science and Engineering, Inc.	Model 1	–	–	3	6 to 11	–	–	–	–
	Model 2000	–	40	–	40	–	40	–	–
Anchoring, Inc.	Model 2-6"	1	5	–	–	–	–	–	–
	Model 3-4"	–	–	1	10	–	–	–	–
	Model 3-6"	3	10	1	10	–	–	–	–

continued

Table A-3. Continued

Agency	Anchor	Soft Clay (Mud)		Medium to Stiff Clay		Sand or Sand/Gravel		Coral and Rock	
		No. of Tests	Penetration (ft)	No. of Tests	Penetration (ft)	No. of Tests	Penetration (ft)	No. of Tests	Penetration (ft)
U.S. Navy (NAVFAC)	Stake Pile								
	Class C 8-in.	2	35	–	–	–	–	–	–
	Class B 12-in.	2	35	–	–	6	34 to 44	–	–
	Class A 16-in.	2	35	–	–	1	35	–	–
	Umbrella Pile								
	Mark III	–	–	–	–	1	19	–	–
	Mark IV	2	13	–	–	2	17 to 18	–	–
U.S. Navy (CB)	Jetted Anchor	–	–	–	–	22	6 to 9	–	–

[a] Magnavox Co. has done extensive testing in simulated laboratory conditions and in on-site situations. The exact number of tests is not known. The figures listed are approximations based on graphs and other data provided by the company.

[b] This low penetration resulted from use of a reduced propellant charge; penetration in excess of 15 feet would be typical.

[c] Original umbrella flukes used.

Appendix B

CURVES FOR SHORT-TERM STATIC
HOLDING CAPACITY VERSUS DEPTH

This appendix presents curves of short-term static holding capacity versus depth for the operative anchors of Chapter 3 when soil properties must be assumed. Figures B-1, B-2, and B-3 show short-term static holding capacity versus depth for small, intermediate, and large anchors, respectively, when they are to be used in the cohesive soil of Figure 5-7. Figures B-4, B-5, and B-6 show short-term static holding capacity versus depth for small, intermediate, and large anchors, respectively, when they are to be used in cohesionless soil where $\phi = 30$ degrees and $\gamma_b = 60$ pcf.

The curves of this appendix also provide a means of comparing the relative holding capabilities of the variety of operative uplift-resisting anchors presented in Chapter 3.

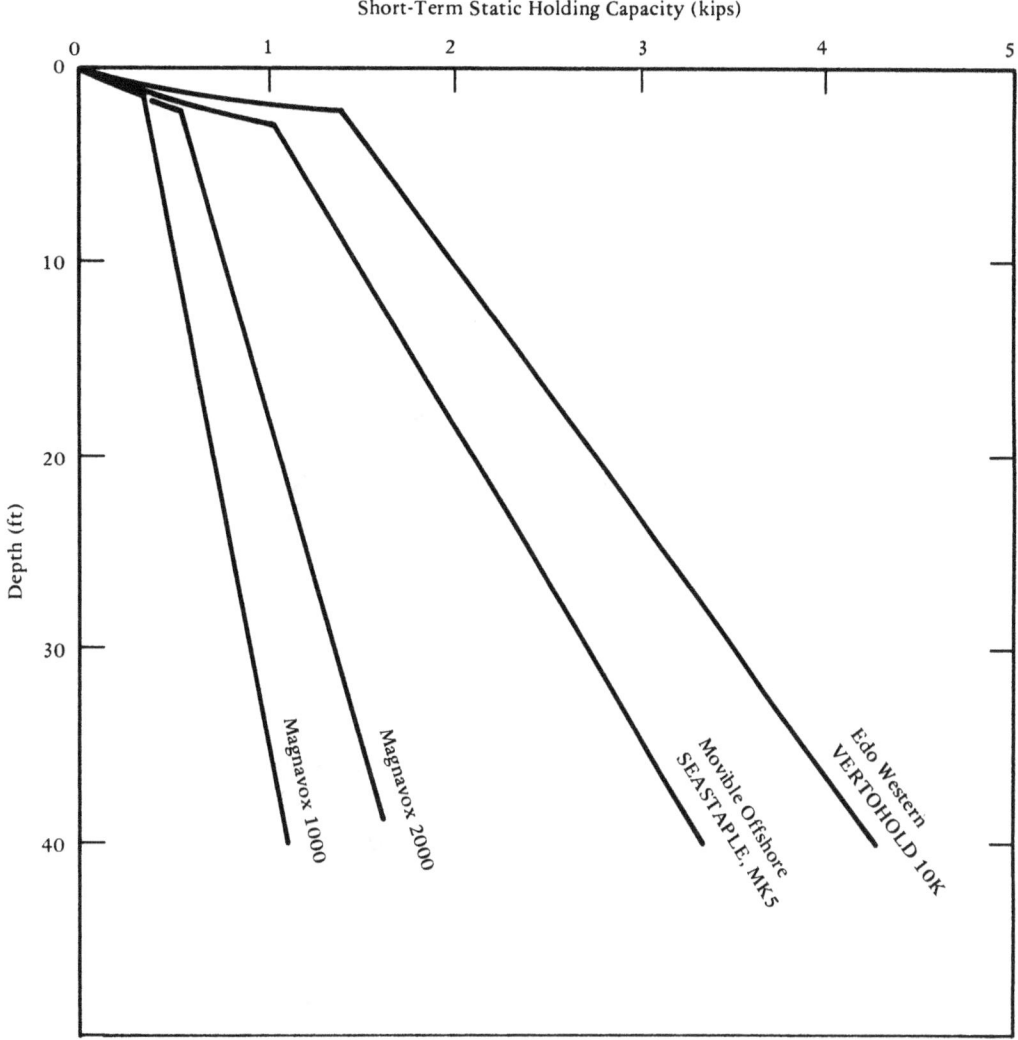

Figure B-1. Short-term static holding capacity versus depth for small uplift-resisting anchors embedded in the cohesive soil described by Figure 5.2-5.

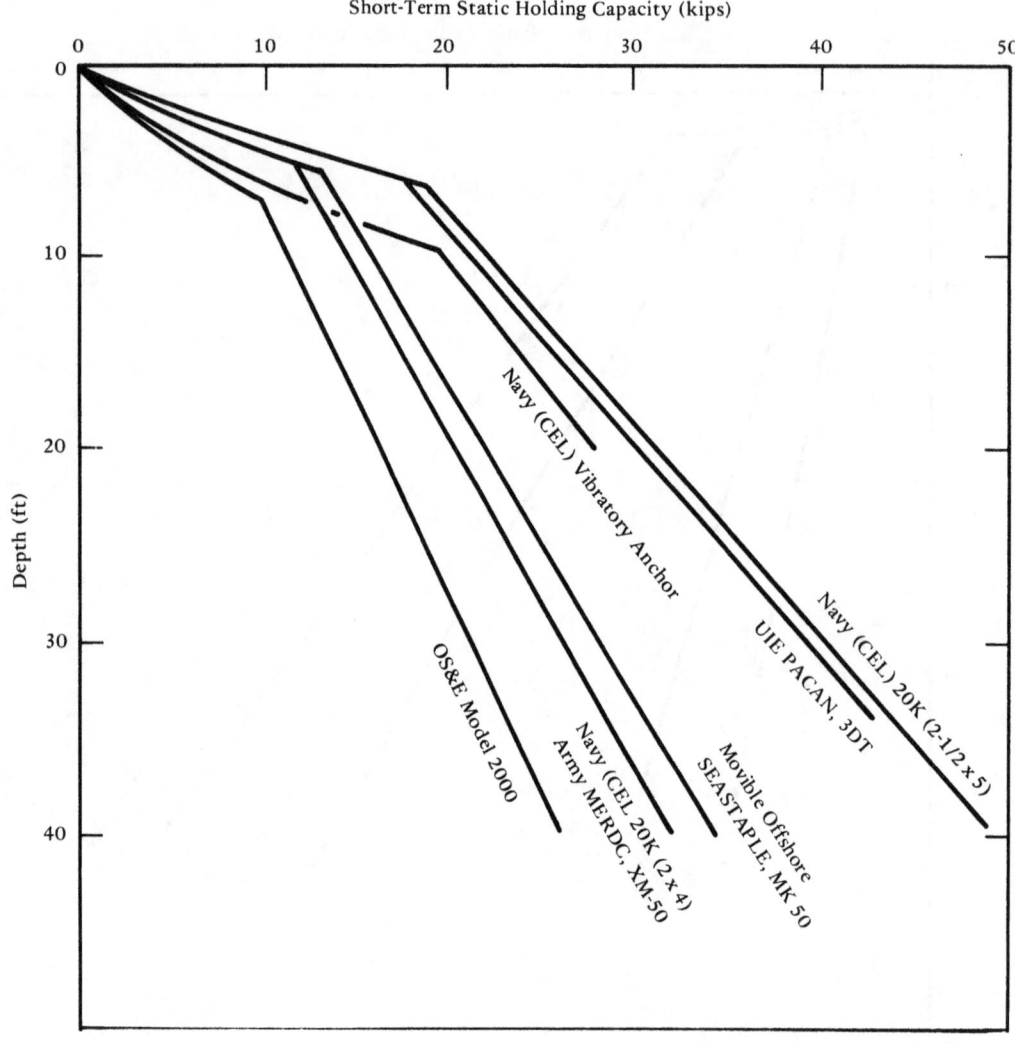

Figure B-2. Short-term static holding capacity versus depth for intermediate uplift-resisting anchors embedded in the cohesive soil described by Figure 5.2-5.

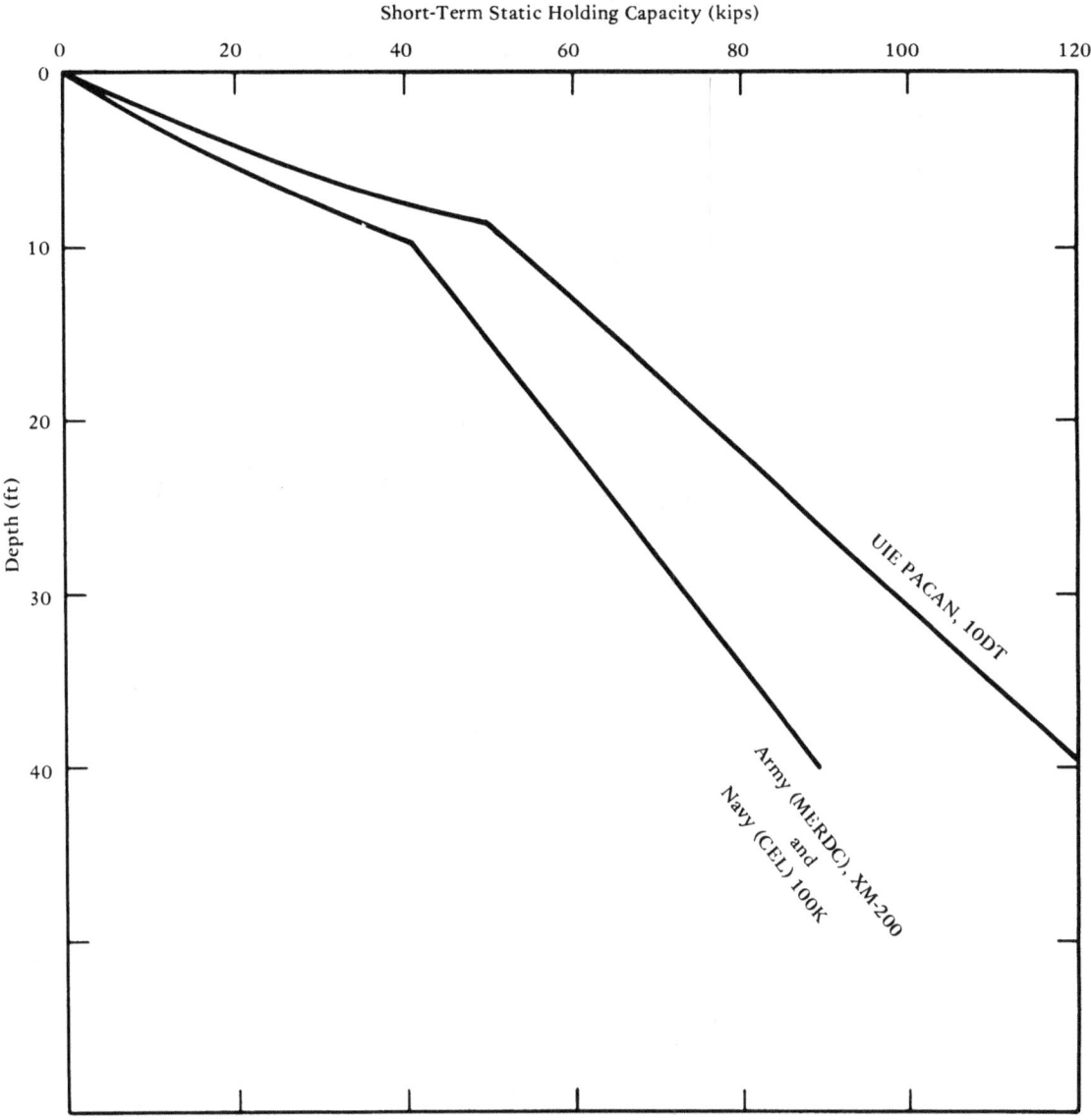

Figure B-3. Short-term static holding capacity versus depth for large uplift-resisting anchors embedded in the cohesive soil described by Figure 5.2-5.

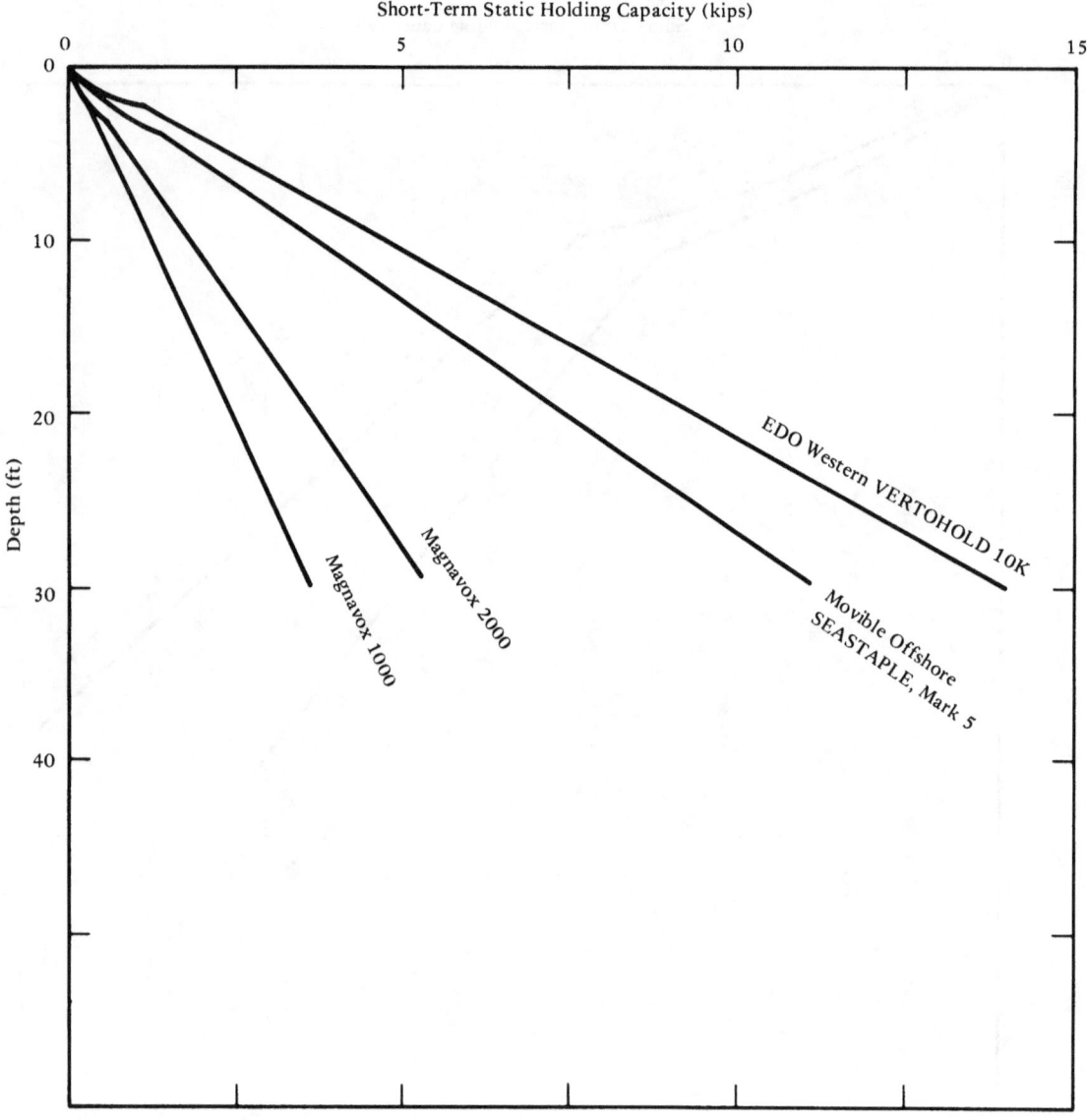

Figure B-4. Holding capacity versus depth for small uplift-resisting anchors embedded in the sand described by $\phi = 30°$ and $\gamma_b = 60$ pcf.

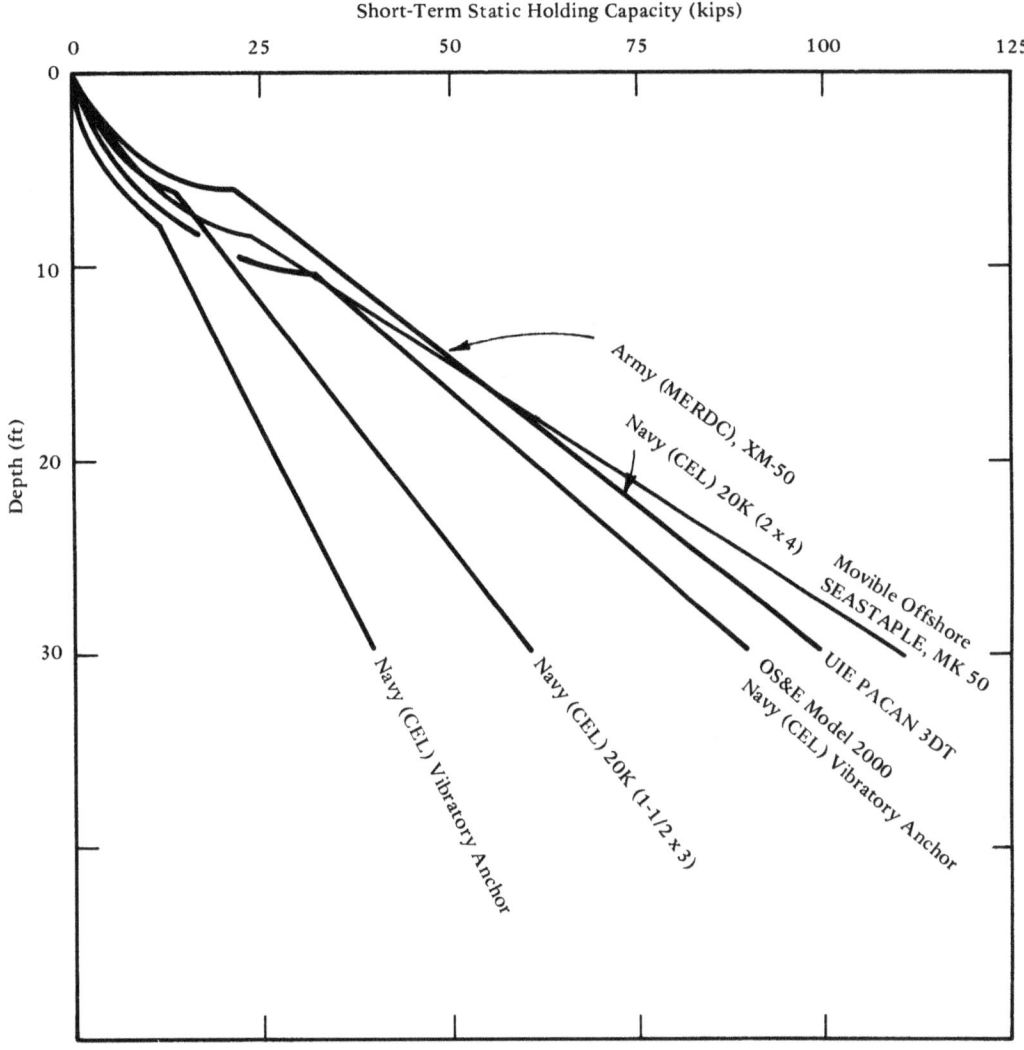

Figure B-5. Holding capacity versus depth for intermediate uplift-resisting anchors embedded in the sand described by $\phi = 30°$ and $\gamma_b = 60$ pcf.

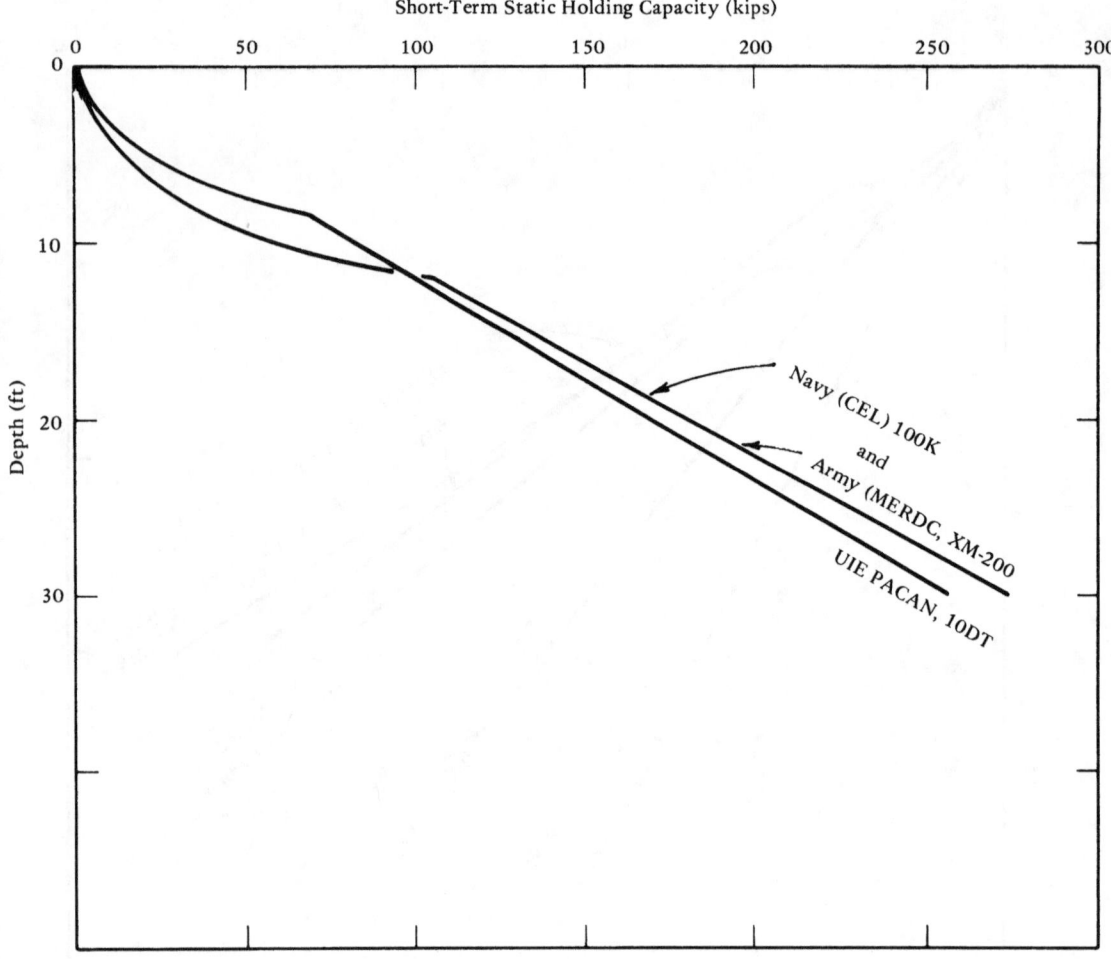

Figure B-6. Holding capacity versus depth for large uplift-resisting anchors embedded in the sand described by $\phi = 30°$ and $\gamma_b = 60$ pcf.

Appendix C

NOMOGRAPHS FOR CALCULATING HOLDING CAPACITY

The nomographs provide an expedient method for solving the basic holding capacity Equation 5-9 in Chapter 5 after the parameters in the equation have been evaluated. Figures C-1, C-2, and C-3 are for calculating the short-term static holding capacity in cohesive soils in the ranges of 0 to 10, 0 to 50, and 0 to 200 kips, respectively. Figures C-4, C-5, and C-6 are for calculating the short-term static holding capacity in cohesionless soils in the ranges of 0 to 10, 0 to 100, and 100 to 300 kips, respectively. A sample problem is presented with each nomograph to illustrate usage of the nomograph.

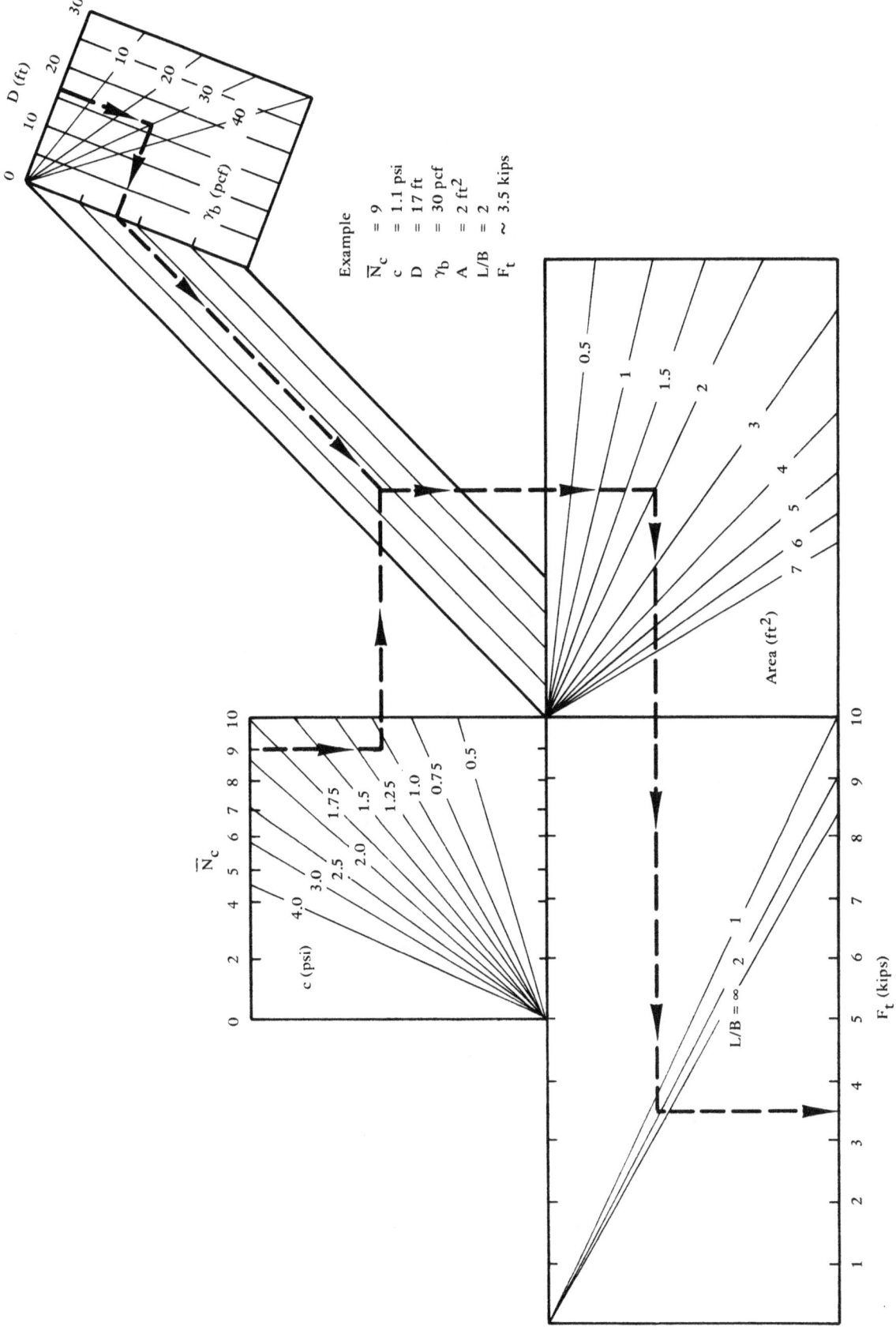

Figure C-1. Nomograph for calculating short-term holding capacity in cohesive soil in the 0-to-10-kip range.

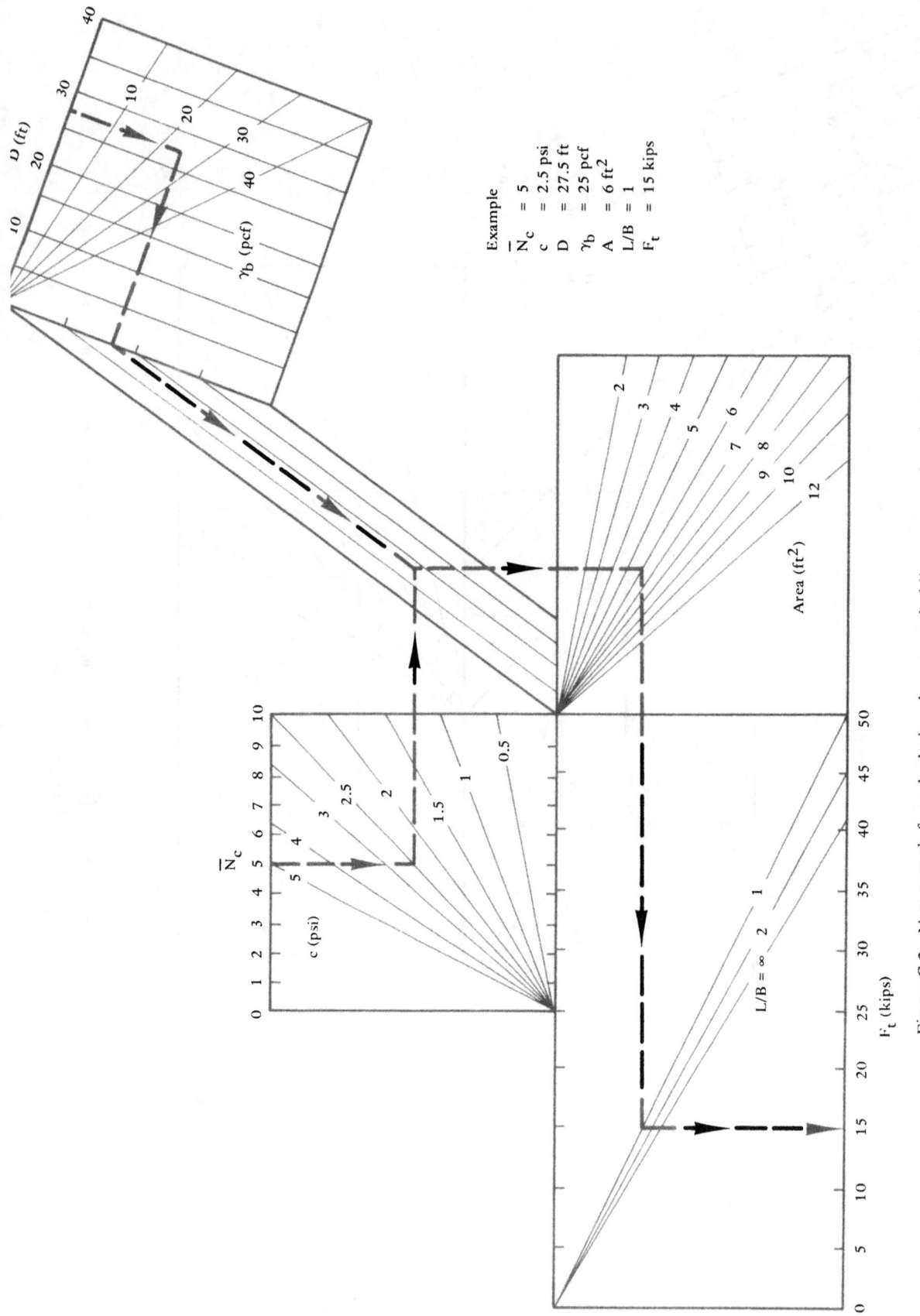

Figure C-2. Nomograph for calculating short-term holding capacity in cohesive soil in the 0-to-50-kip range.

149

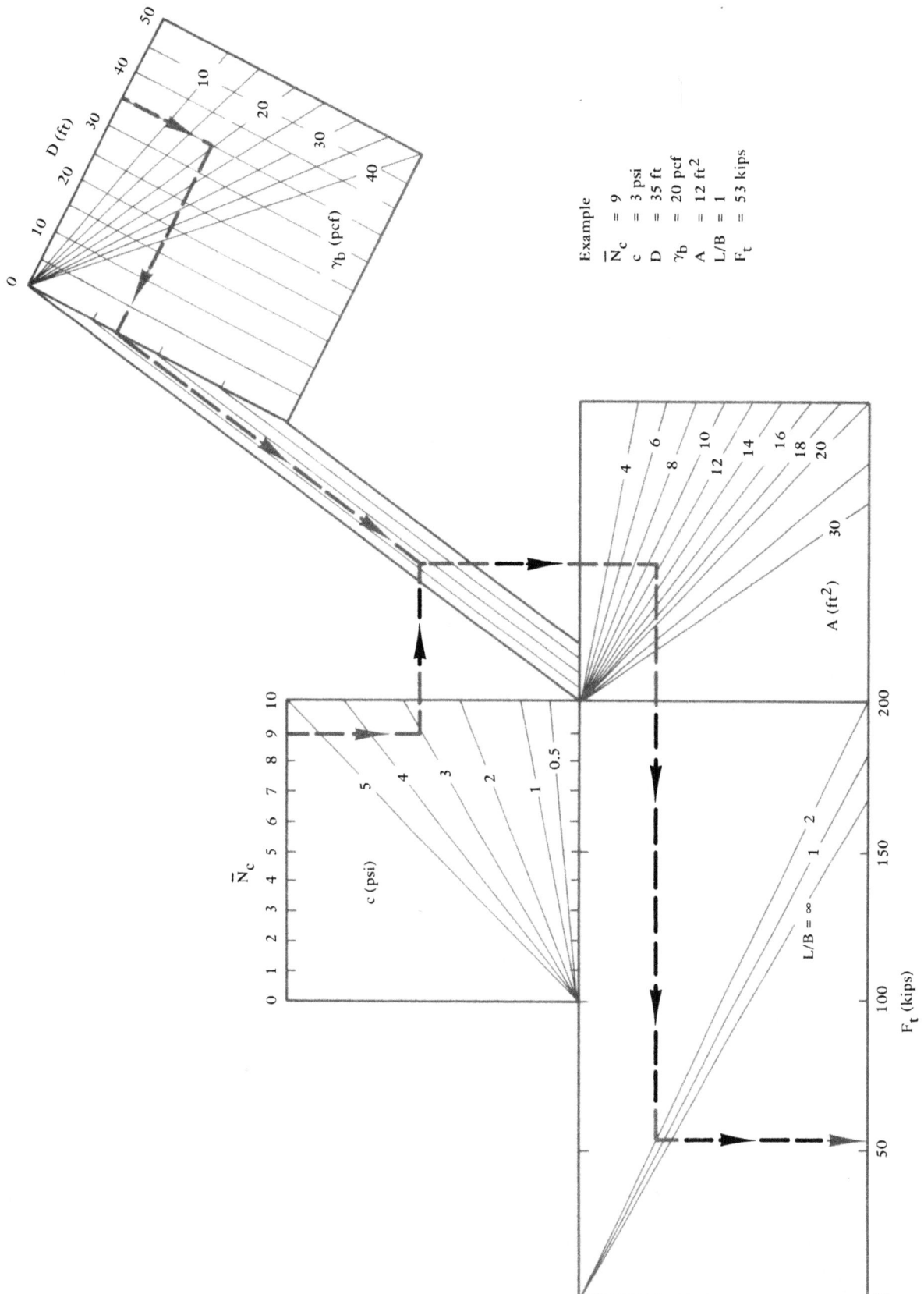

Figure C-3. Nomograph for calculating short-term holding capacity in cohesive soil in the 0-to-200-kip range.

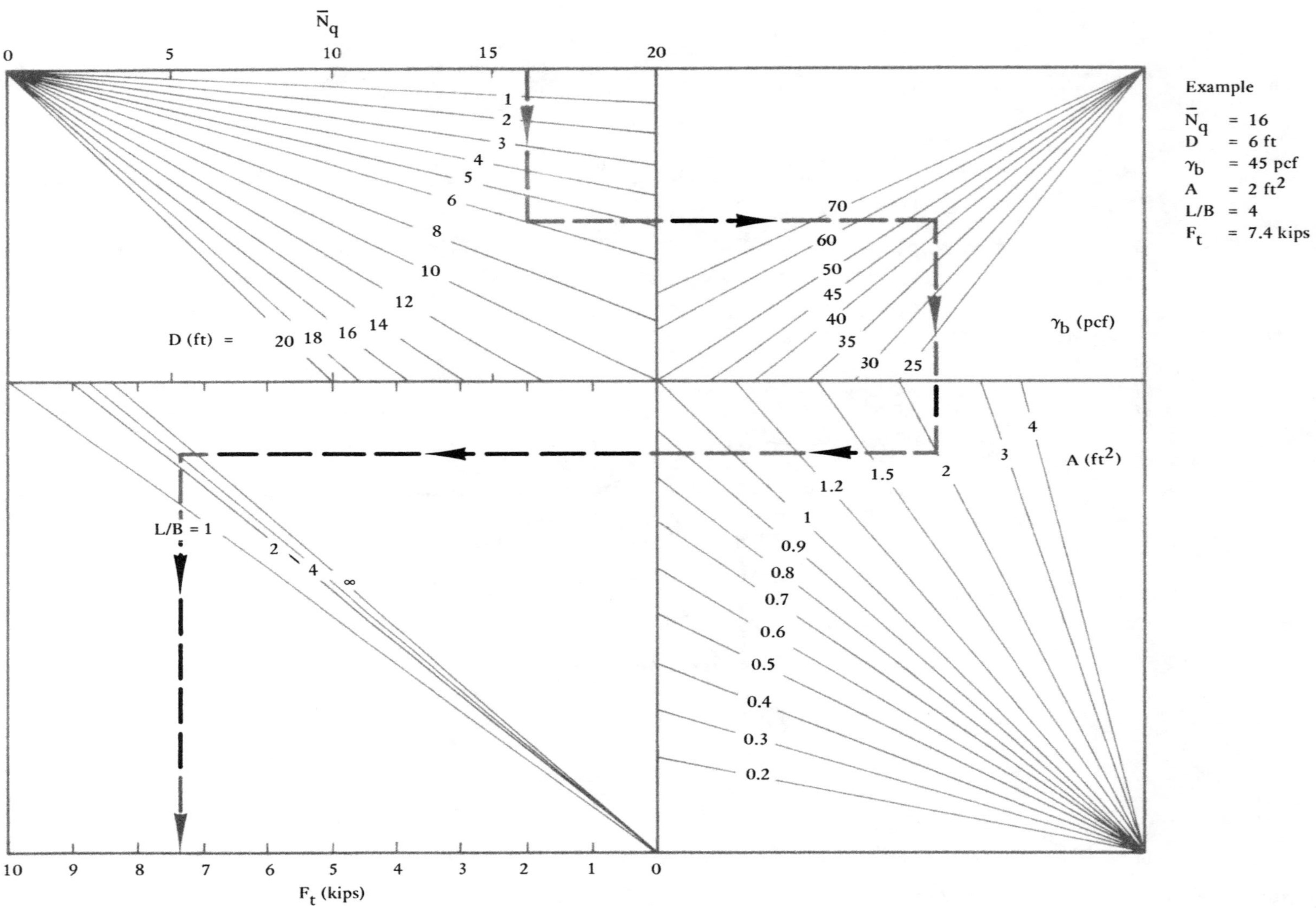

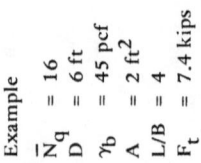

Example
$\bar{N}_q$ = 16
D = 6 ft
γ_b = 45 pcf
A = 2 ft²
L/B = 4
F_t = 7.4 kips

Figure C-4. Nomograph for calculating holding capacity in sand in the 0-to-10-kip range.

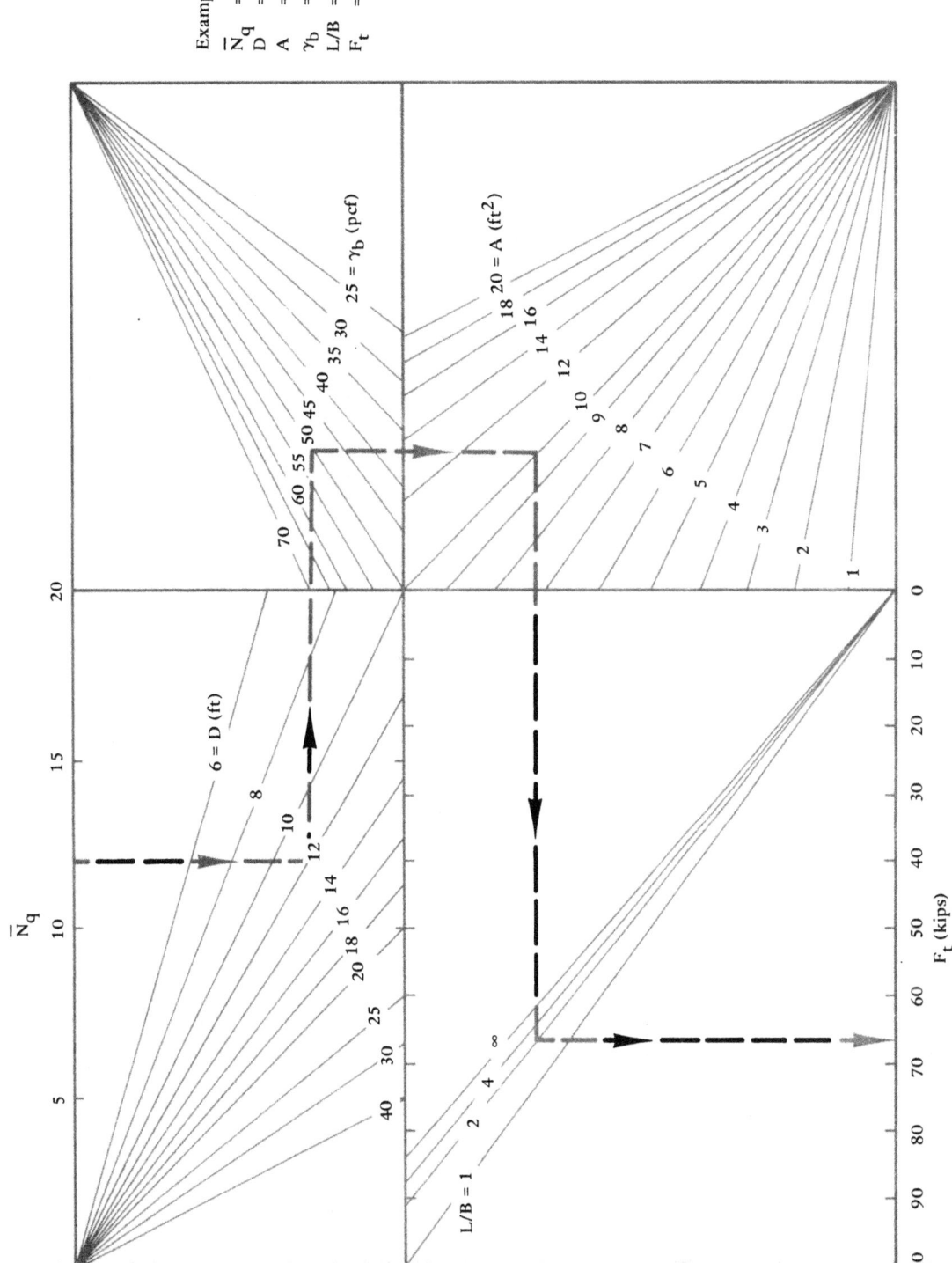

Figure C-5. Nomograph for calculating holding capacity in sand in the 0-to-100-kip range.

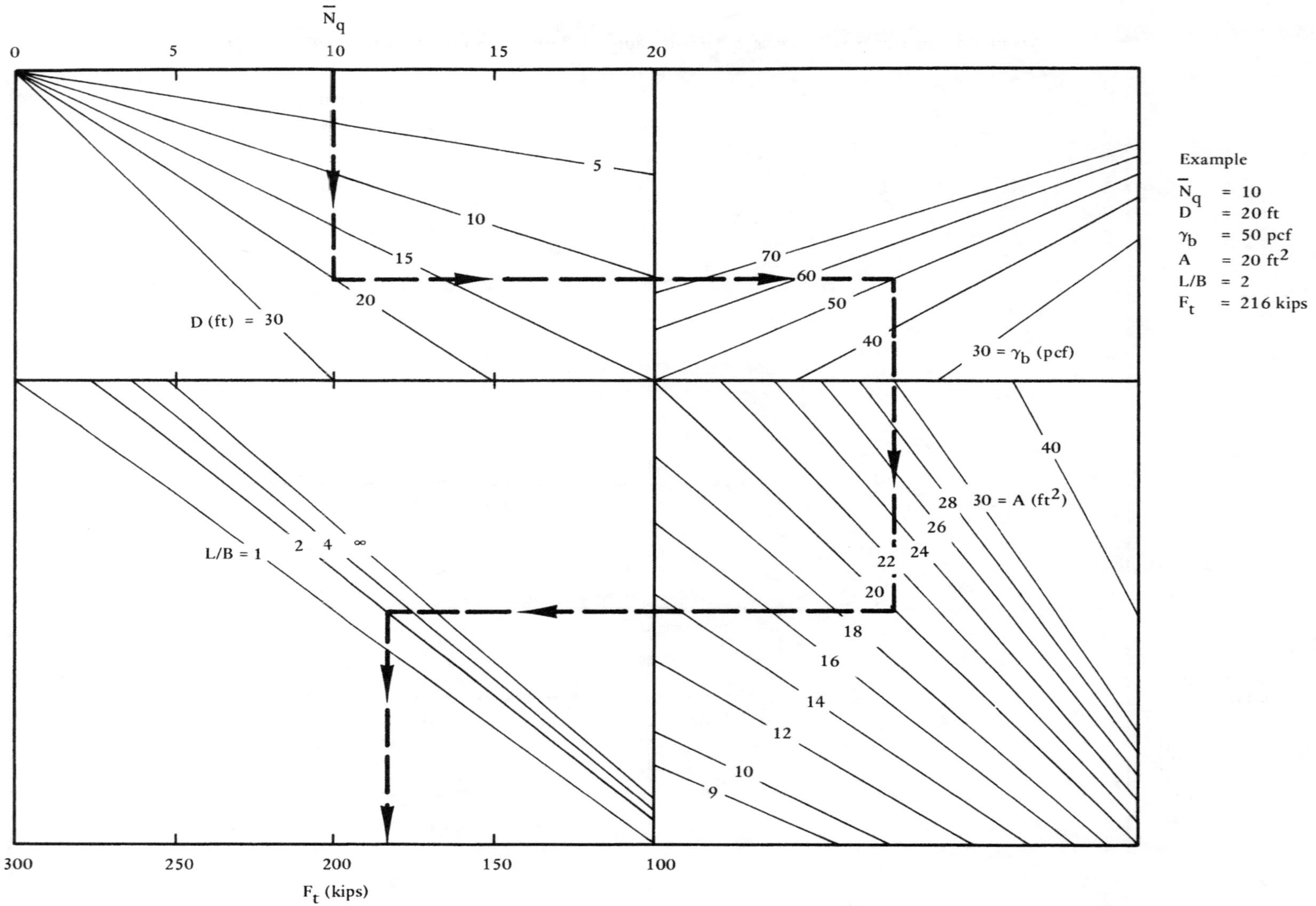

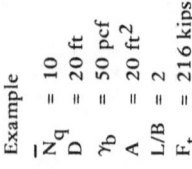

Figure C-6. Nomograph for calculating holding capacity in sand in the 0-to-300-kip range.

www.ingramcontent.com/pod-product-compliance
Lightning Source LLC
Chambersburg PA
CBHW082123230426
43671CB00015B/2784